ENVIRONMENTAL SCIENCE, ENGINEERING AND TECHNOLOGY SERIES

ENVIRONMENTAL MODELING WITH GPS

Environmental Science, Engineering and Technology Series

Nitrous Oxide Emissions Research Progress
Adam I. Sheldon and Edward P. Barnhart (Editors)
2009. ISBN: 978-1-60692-267-5

Fundamentals and Applications of Biosorption Isotherms, Kinetics and Thermodynamics
Yu Liu and Jianlong Wang (Editors)
2009. ISBN: 978-1-60741-169-7

Environmental Effects of Off-Highway Vehicles
Douglas S. Ouren, Christopher Haas, Cynthia P. Melcher, Susan C. Stewart, Phadrea D. Ponds, Natalie R. Sexton, Lucy Burris, Tammy Fancher and Zachary H. Bowen
2009. ISBN: 978-1-60692-936-0

Agricultural Runoff, Coastal Engineering and Flooding
Christopher A. Hudspeth and Timothy E. Reeve (Editors)
2009. ISBN: 978-1-60741-097-3

Agricultural Runoff, Coastal Engineering and Flooding
Christopher A. Hudspeth and Timothy E. Reeve (Editors)
2009. ISBN: 978-1-60876-608-6 (Online Book)

Conservation of Natural Resources
Nikolas J. Kudrow (Editor)
2009. ISBN: 978-1-60741-178-9

Conservation of Natural Resources
Nikolas J. Kudrow (Editor)
2009. ISBN: 978-1-60876-642-6 (Online Book)

Directory of Conservation Funding Sources for Developing Countries: Conservation Biology, Education and Training, Fellowships and Scholarships
Alfred O. Owino and Joseph O. Oyugi
2009. ISBN: 978-1-60741-367-7

Forest Canopies: Forest Production, Ecosystem Health and Climate Conditions
Jason D. Creighton and Paul J. Roney (Editors)
2009. ISBN: 978-1-60741-457-5

Soil Fertility
Derek P. Lucero and Joseph E. Boggs (Editors)
2009. ISBN: 978-1-60741-466-7

The Amazon Gold Rush and Environmental Mercury Contamination
Daniel Marcos Bonotto and Ene Glória da Silveira
2009. ISBN: 978-1-60741-609-8

Process Engineering in Plant-Based Products
Hongzhang Chen
2009. ISBN: 978-1-60741-962-4

Buildings and the Environment
Jonas Nemecek and Patrik Schulz (Editors)
2009. ISBN: 978-1-60876-128-9

Tree Growth: Influences, Layers and Types
Wesley P. Karam (Editor)
2009. ISBN: 978-1-60741-784-2

Syngas: Production Methods, Post Treatment and Economics
Adorjan Kurucz and Izsak Bencik (Editors)
2009. ISBN: 978-1-60741-841-2

Syngas: Production Methods, Post Treatment and Economics
Adorjan Kurucz and Izsak Bencik (Editors)
2009. ISBN: 978-1-61668-214-9 (Online Book)

Carbon Capture and Storage including Coal-Fired Power Plants
Todd P. Carington (Editor)
2009. ISBN: 978-1-60741-196-3

Process Engineering in Plant-Based Products
Hongzhang Chen
2009. ISBN: 978-1-60741-962-4

Potential of Activated Sludge Utilization
Xiaoyi Yang
2010. ISBN: 978-1-60876-019-0

Psychological Approaches to Sustainability: Current Trends in Theory, Research and Applications
Victor Corral-Verdugo, Cirilo H. Garcia-Cadena and Martha Frias-Armenta (Editors)
2010. ISBN: 978-1-60876-356-6

Fluid Waste Disposal
Kay W. Canton (Editor)
2010. ISBN: 978-1-60741-915-0

Recent Progress on Earthquake Geology
Pierpaolo Guarnieri (Editor)
2010. ISBN: 978-1-60876-147-0

Check Dams, Morphological Adjustments and Erosion Control in Torrential Streams
Carmelo Consesa Garcia and Mario Aristide Lenzi (Editors)
2010. ISBN: 978-1-60876-146-3

Freshwater Ecosystems and Aquaculture Research
Felice De Carlo and Alessio Bassano (Editors)
2010. ISBN: 978-1-60741-707-1

Grassland Biodiversity: Habitat Types, Ecological Processes and Environmental Impacts
Johan Runas and Theodor Dahlgren (Editors)
2010. ISBN: 978-1-60876-542-3

Handbook on Agroforestry: Management Practices and Environmental Impact
Lawrence R. Kellimore (Editor)
2010. ISBN: 978-1-60876-359-7

Biodiversity Hotspots
Vittore Rescigno and Savario Maletta (Editors)
2010. ISBN: 978-1-60876-458-7

Handbook of Environmental Policy
Johannes Meijer and Arjan der Berg (Editors)
2010. ISBN: 978-1-60741-635-7

Environmental Modeling with GPS
Lubos Matejicek (Editor)
2010. ISBN: 978-1-60876-363-4

ENVIRONMENTAL SCIENCE, ENGINEERING AND TECHNOLOGY SERIES

ENVIRONMENTAL MODELING WITH GPS

LUBOS MATEJICEK
EDITOR

Nova Science Publishers, Inc.
New York

For permission to use material from this book please contact us:
Telephone 631-231-7269; Fax 631-231-8175
Web Site: http://www.novapublishers.com

LIBRARY OF CONGRESS CATALOGING-IN-PUBLICATION DATA

Available upon request.

ISBN: 978-1-60876-363-4

Published by Nova Science Publishers, Inc. ✢ New York

CONTENTS

PREFACE

Description of principles focused on environmental interactions can be developed in a more sophistic way by spatio-temporal models that are able to manage spatial data together with temporal data. A software tool offering data management with spatial and time dependent attributes is represented by Geographic Information System (GIS). Whereas the GISs manage spatial description of the environment, dynamic modeling is focused on functioning of environmental processes, such as environmental pollution or population dynamics. In spite of that the dynamic modeling and GIS are synergistic, both systems grew up separately. The roots of dynamic modeling come mostly from physics, which offers a wide range of deterministic interactions described by physical laws and rules. The GISs originate from cartographic endeavour to automate map development and updates. But nowadays, both these individual research areas are linked together with their possibilities to support description of spatial interaction and time phenomena in the frame of environmental models. For all that, the computer programs, which represent dynamic modeling and GIS, have different data structures and functions for data management and numerical calculations. This makes it difficult to link inputs and outputs together to carry out complex analysis.

In order to study individual factors of pollution and stress, environmental protection includes a wide variety of techniques that overtax many of present data structures. Thus, new concepts of data links and sharing of functions are needed. In the past, air pollution, contamination of soils and water, noise assessment, waste management and landscape protection were monitored in the local scale. Independent computer tools were used to analyze data measurements and to predict effects. At present, environmental complex analysis of urban and rural areas dictates integration of the individual dynamic models and data structures into a complex system. A number of case oriented solutions are described in the book edited by Goodchild et al. (1996) "GIS and Environmental Modeling: Progress and Research Issues".

Environmental modeling contains a wide range of experimental research activities. The main aims represent estimates of short term and long term changes in various scales, data analysis, and development of models that can simulate real situation to aid decision-making processes. Due to a huge number of relatively different research methods, the aim of this book is focused on environmental pollution that is used to be divided into individual disciplines. Generally, modeling of air pollution includes localization of global and local sources, description of particle transport and chemical interaction in atmosphere. Development of surface water and groundwater models together with soil pollution is focused on similar phenomena but in the different physical and chemical environment. In the framework of environmental modeling, the dispersion of pollutants is mostly considered as a combination of accumulation and transfer phenomena in the abiotic parts of ecological systems. Biotic parts can be affected consecutively. The description of spreading pollutant in dynamic models has to be subordinated to ecological interactions. Complex studies of the living environment should be also complemented by the assessment of landscape protection, waste management, noise assessment and other environmental impacts that affect the living environment, such as biomonitoring, allergy assessment, etc.

Air pollution is usually explored by monitoring of pollutants over urban sites and rural areas, which belongs to the standard routines of environmental assessment. The systems are usually divided on registering of air pollution sources, and assessment of air quality on the basis of monitoring and other relevant information (smog regulation system, meteorological measurements and forecast). The sources of atmospheric pollution are used to be registered into a few categories that include stationary sources and mobile sources. The assessment of air quality is based on the air quality limits for pollutants, which can be related to average annual, daily, 8-hour's and half-hour pollutant concentration. The assessment of air pollution situation is based on data stored in central databases. Mostly, the monitored pollutants are represented by sulphur dioxide, nitrogen oxides, ground-level ozone, carbon monoxide, suspended particulate matter and aromatic hydrocarbons. Other information about air pollution can be given by atmospheric deposits, dust fallout, heavy metals and organic compounds. A wide range of data and measurements from air pollution monitoring networks has to be stored and effectively processed on computer information systems. Evaluation of the actual situation and the long term trends of air pollution require integration of a wide range of data, usually in the form of time series together with geographical information about the location of pollution sources, monitoring sites, and residential, industrial and traffic areas. The primary data have to be processed and analyzed in order to derive specific characteristics for decision-making processes. The basic level covers geostatistical procedures such as spatial correlation and regression, factor analysis and spatial interpolation. In the framework of GISs, the map algebra can be applied to estimate more complex spatio-temporal characteristics for risk assessment. The superstructure is represented by dynamic models that are focused on transport of air pollution and possibly chemical interactions in the living environment. Whereas the GISs are used for data management and geostatistical analysis, the dynamic modeling of air pollution is mostly processed with standalone computer programs that include implementations of numerical algorithms for solution of partial differential equations. In spite of a wide range of functions, which are implemented in the GIS for spatial data management, existing GISs do not offer effective programming environment for implementation of time and capacity consuming calculations for dynamic modeling. On the opposite site, the GISs offer far more functions for storage and visualization of spatio-temporal data. Thus, the recent GISs are mostly used for pre-processing (data input, basic data analysis and estimation of model parameters) and post-processing (data management and visualization, printing maps and presentation of results on the Internet).

Water pollution is linked with aquatic ecosystems that perform numerous valuable environmental functions. They cycle nutrients, purify water, attenuate floods, recharge ground water and provide habitat for wildlife and recreation. Increased population accompanied by intensified industrial, commercial, and resident development has caused pollution of water by fertilizers, insecticides, oil, toxic, and feedlot waste. Water pollution is used to be divided into characteristic groups according to the type of pollution (organic compounds, heavy metals, thermal pollution and radioactivity) and the author of pollution (sewage from people, agricultural, industry). Organic pollution occurs when large quantities of organic compounds are released into watercourse. During the decomposition processes, the dissolved oxygen is reduced, which causes the oxygen depletion. Heavy metals have a number of chemical properties in common. Their biological roles range from the essential to the highly toxic. For example, mercury, cadmium and lead have no biological role in the healthy organism, but can cause a serious damage even at very low concentration levels. Many heavy metals are accumulated in organisms. Thermal pollution is mostly caused by electricity-generating plants that operate through the thermodynamic process, in which the high pressure steam is produced and then expanded by turbines. The condensers require large amount of cooling water, which is removed from ground and surface waters and returned at a higher temperature. In spite of that the released water is mostly free of pollutants, the increase of temperature is a very serious problem. It alters the physical environment in terms of both the reduction in the water density, and in its oxygen concentration, which varies inversely to with the temperature. Radioactivity means the phenomenon in which the nucleus spontaneously emits radiation. The environment is naturally radioactive, with radiation coming from the outer space, from the earth, from the atoms within the body, and from other activities, such as burning fuel and cultivating soil. The major man-made sources of radiation to the living environment are those from nuclear weapons testing, and from nuclear industry. The main concern over radioactivity is obviously the contamination of the food chain, leading to the man. The normal wastes released from nuclear power plants have not been shown to cause harm to the living environment or to humans. In the framework of water pollution, water monitoring is recommended to identify abnormal concentration, to define the peaks, to estimate mean concentrations, to detect trends or changes. All these objectives can be summarized in order to assess the state of rivers and the variability of quality, to determine action if necessary, to maintain and enhance quality, and to assess the effectiveness of such actions.

Soil pollution is mostly related to the places exposed by industrial and agricultural activities. Whereas some parameters of air and water pollution are obtained from automatic and manually-operated monitoring networks, the levels of soil pollution are estimated from individual samples. To preserve comparability, the samples are collected from the same places. Examination of long-term changes requires expensive laboratory analyses. Regular measurements are usually complemented by data obtained from the local environmental project focused on dispose of soil contamination by disasters and waste dumps. The principles of evaluation are based on rules carried out by the local and state authorities. They determine the maximum permissible limits of hazardous elements (heavy metals, inorganic and organic substances, polycyclic aromatic hydrocarbons). In spite of that the primary measured data do not require building of a large database structure, they represent key factors for complex model analysis and risk assessment. Spatial data structures supported by the GIS together with measured data provide the base for geostatistical methods and dynamic modeling. The geostatistical methods are focused on spatial interpolations based on terrain measurements. The analysis can be extended by using data from digital elevation models (DEM), and by using functions derived from map algebra. Dynamic modeling of soil pollution is mostly focused on accumulation, transport and interaction of contaminants in soils. Due the heterogeneous environment, the model description is often represented by the partial differential equations with spatially dependent parameters. But the lack of data to estimate all the parameters and initial conditions must be mostly eliminated by simplified model description. The GIS is often used for pre-processing and post-processing routines, but the solution of simplified models can be carried out in the GIS programming environment.

Landscape protection represents other important element. Due to the violation of former ecological systems and heavy antropogenic impacts, extra funds and remediation projects are necessary to preserve existing nature. Conservation of flora and fauna include mapping of the existing conditions and monitoring short-term and long-term trends. Information systems support the evidence of land-use and land-cover. Besides the cadastral maps, information about protected areas and natural parks involve other data sources for decision-making management. GIS together with remote sensing can map actual situation. Aerial photographs and satellite images help to identify and classify changes of the landscape morphology. Then, all the data can be processed by the GIS. From the modeling perspective, landform changes can be explored by a wide range of GIS tools. In comparison with modeling of pollution, the methodology and modeling tools are focused on observation of complex natural, cultural and historical characteristics of landscape.

Waste management focused on the waste treatment and waste disposal has been improved from its early beginnings of mere dumping to a sophisticated range of options including today operations, such as recycling, incineration with energy recovery, advanced landfill design, and alternative technology. Integrated waste management has been defined as the integration of waste stream, collection and treatment methods, environmental benefit, economic optimization and social acceptability into a practical system. Wastes are divided into many categories, such as household waste, industrial waste, clinical waste and municipal solid waste. Other special category is formed by hazardous wastes that are dangerous and difficult to treat, keep or dispose. Thus a special provision is required for dealing with them. They may contain substances that are corrosive, toxic, reactive, infectious, carcinogenic or harmful to human health, and living environment.

Noise assessment is focused on excessive noise and its long-term effects. A number of ailments (stress, neuroses, undesirable blood pressure changes) are manifested with a delay of several years. Information about noise levels, which are expressed by a number of parameters (equivalent noise level A LAeq for various built-up areas, daytime and nightime; maximum and minimum noise levels), are obtained by measurements and models. The final estimates are put to the noise maps that are used for decision-making processes of endorsement and sanctioning. A wide range of information about noise sources, urban environment and measurements has to be organized to carry out maps and models. However GISs offer a wide range of functions and spatial data management, many of data and models are still managed in standalone systems. Nowadays, GIS development tools offer improvement of data integration and calculation methods.

Other pollution impacts include many of other components that affect inhabitants in our living environment. Some of them are included in various biomonitoring studies (occurrence of plant that affect allergy, monitoring of occurrences of Lyme boreliosis germs in ticks, occurrences of small rodents, accumulation of heavy metals and organic compounds in food chains). Important data sources represent the case oriented research reports based on specific projects supported by state and industrial authorities. The GIS by itself plays an integration link to a huge number of data formats and calculation tools.

Identification methods for environmental modeling include building of models by using natural laws or empirical experience and estimating their parameters. Models, which can take advantage of natural laws, describe processes more comprehensibly. Considering to a complexity of environmental pollution, the base of models is described with the using of physical laws (transport and accumulation of pollutants). Empirical rules and observations are used to elaborate a number of other phenomena (heterogeneity of urban environment, antropogenic interaction). There are no strict rules for building of models in dynamic modeling, which means that a few different models can be developed for modeling complex processes of environmental pollution. The next phase of identification includes setting of the parameters in mathematical equations. Both these phases need long-term observations of real processes. During the identification and parameter settings, the behavior of the model and the measured data are repeatedly compared. The additional corrections are made to decrease the deviations between the model outputs and the data of a real system.

The GISs also contain a number of identification functions that are mostly focused on estimates of geostatistical models. Spatial interpolations use deterministic and statistical estimates of parameters. Regression and correlation models can operate in addition with spatial data. In spite of these tools, dynamic modeling of environmental pollution contains more comprehensive extent of functionality. The identification methods can be built in the GIS in case of smaller models. The larger the model the more specific methods are needed. In practice, there are not any general identification technique of the model structure and its parameters in the area of environmental modeling. Mostly, the identification is based on user's experience, and the GIS power to integrate a wide range of data and collaborative functionality.

Simulation is used to solve the models represented by mathematical equations that describe fundamental interactions of real processes. The examination of their functionality is carried out by a numerical calculation of the mathematical expressions on digital computers. A number of programming tools and programs have been developed during the last decades to simplify the entry of mathematical expressions and their solutions. For example, Advanced Continuous Simulation Language (ACSL) represents a system for numerical calculation of dynamic models. The dynamics of the model is calculated by a few basic elementary operations: integration and time delay that are complemented with a number of algebraic relations. Especially, integration is used to realize accumulation phenomena in dynamic models. Time delay can support modeling of transport phenomena. Besides the ACSL, which represents the standard tool for calculation of dynamic models, there are many other systems or languages that can carry out similar operations. They are grouped into many classes along the modeling branches (air pollution, soil and groundwater contamination, noise assessment, waste management). The individual software systems use various data formats that cause obstacles for data exchange and complex analysis. The recent systems represent complex tools that can support, for example, groundwater modeling (GMS- Groundwater Modeling System supporting both finite-difference and finite-element models in 2D and 3D including MODFLOW, MODPATH, MT3DMS/RT3D, SEEP2D, SEAM3D, UTCHEM and FEMEWATER), watershed modeling (WMS- Watershed Modeling System- for all phases of watershed modeling inclusive HEC-1, TR-20 and TR-55). Other examples could be illustrated by modeling systems for air pollution, noise assessment, landscape protection, ecological analysis or ecological modeling. Individual group can be formed by environmental models developed in the environment of the MATLAB, Stella, Mathematica or other tools.

In addition to the standard simulation systems, which support data exchange on the level of file sharing or database connection, the GISs operate over a number of formats that support a region of space and time within which the variables can be simulated. Naturally, the GISs are focused mostly on spatial analysis, which can cause incompatibility with a number of standard simulation software tools. Thus, the data have to be often converted by case oriented software extensions.

The dynamic models can be divided into a number of groups (continuous or discrete in space or time, linear or non-linear, deterministic or stochastic). Each group requires different data structures and algorithms for calculation. The GISs supports partially simulation of some dynamic models. For example, the GIS ArcGIS can solve a number of model phenomena, such as diffusion and transport functions. The GIS Idrisi is extended with a number of modules that support dynamic control structures allowing the development of temporal models, such as cellular automata, growth and change analysis. Besides standard functions, almost each GIS contains programming tools that enable implementation of own numerical algorithms for calculation of environmental models. Because more complex models have considerable demands on efficiency of numeric calculations, some dynamic models are used to be solved outside of the GIS environment.

Many simulation systems are focused on time consuming numerical calculations, but visualization is only supported on the basic level. Besides, each system contains specific functions and a different user environment. The GISs offer a standard environment and the possibility to extend graphic outputs with spatial data, images and photos. In addition, all the graphic outputs can be accessible through the Internet by the GIS browsers, such as ArcExplorer and CAD tools.

In environmental modeling, the spatio-temporal models are based on mathematical expressions that represent interactions of environmental processes. During the validation and verification, the structure and measured characteristics of the real system are repeatedly compared with the model structure and simulation outputs. In GIS, the abstract model rests mainly on the representation of geographic features by data connected to layers of points, lines, polygons, grid cells, images and photos. The processing of subsequent data with functions represents next steps that is mostly focused on spatial interaction and transformation of the original objects. In spite of that verification and validation are terms used in the context of dynamic modeling, the evaluation procedures in the GIS environment mean how accurate and complete spatial data are (location, shape, generalization and topology).

Verification and validation from the perspective of environmental dynamic models built inside the GIS associate both the time and spatial aspects. Especially in the urban areas, the complex spatio-temporal analyses represent multilateral problems. Considering to the heterogeneity of modeling systems, verification and validation are often pursued separately for individual components of environmental pollution. Unfortunately, the multidiscipline orientation of the complex environmental models overtaxes the area of individual research specializations (air, water and soil pollution; landscape protection; noise assessment; GIS; dynamic modeling), which opens new research ways.

The difference between the spatial objects and the time dependent phenomena of real systems has been caused distinct approaches in error assessment. The dynamic models generalize real systems through the fundamental and insignificant phenomena relative to the studied subjects. In accordance with this, discrete or continuous, and linear or non-linear models are used. The spatial analysis generalizes the real object in view of the scale.

Similarly, this indicates the use of the vector or raster spatial data formats, pertinently triangulated irregular networks (TIN) for digital elevation models or other spatial formats. Despite of various levels of the data accuracy, the complex analysis needs compatible data formats to analyze interactions of environmental pollution together with the living environment. One of the ways is represented by cell based modeling that can assess each cell in the grid by weighted factors of individual environmental processes. Data flows start from data collection, model development and basic evaluation to the model simulation and the selected methods of statistical analysis. The final analyses are carried out in the GIS. In addition to a number of spatial analyses, the cell based modeling can be used for complex analysis of individual data outputs. In spite of the transparent data management in the form of the reclassified regular grids, the dynamic phenomena of environmental pollution can be studied by a restricted way. Thus, the cell based modeling, which is mostly used for basic data assessment, cannot provide more complex time dependent analyses that are carried out by simulation of individual dynamic models.

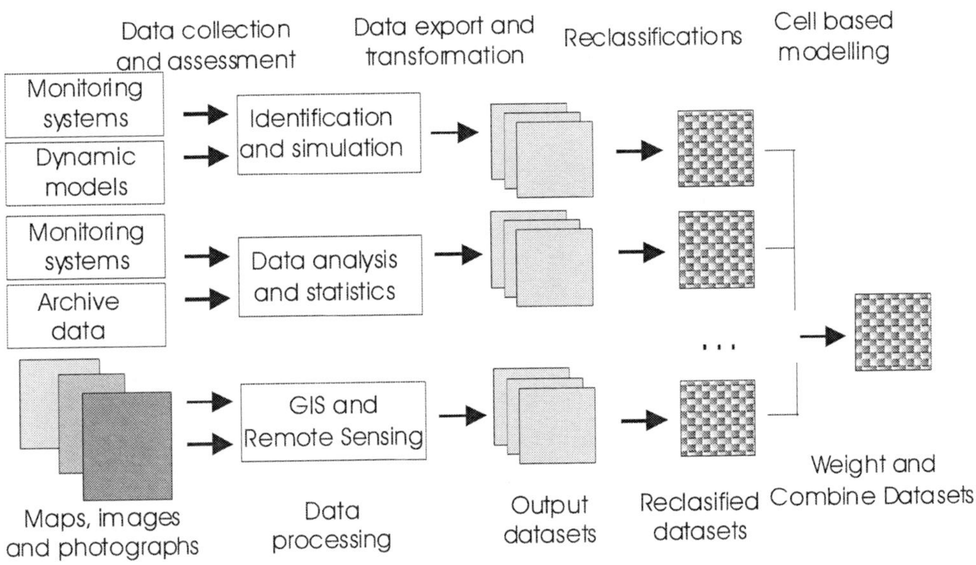

Cell based modelling focused on environmental pollution in urban sites and rural areas.

In: Environmental Modeling with GIS
Editor: Lubos Matejicek, pp. 1-15

ISBN: 978-1-60876-363-4

Chapter 1

SPATIO-TEMPORAL MODELING OF RADIONUCLIDE DEPOSITION

Lubos Matejicek
Institute for Environmental Studies, Charles University in Prague, Faculty of Natural Science, Benatska 2, 128 01, Prague, Czech Republic

ABSTRACT

An integrated research on spatio-temporal modeling of radionuclide deposition is solved by a standard way, and by advanced GIS tools. In order to explore radionuclide transport in the Irish Sea, a dynamic compartment model is developed for accumulation and redistribution of plutonium ($^{239/240}Pu$) over a few decades. The compartments are situated in regions that are divided into a few vertical parts: the water column, the interface and the sediment parts. Basic transport processes are represented by advection and diffusion. Advection denotes transport by shifting particles from one place to another by the flow field. Diffusion originates from concentration differences that result in a tendency to equalize concentration levels. In dependence on the environment, there are different diffusion speeds in surface water bodies and in sediments, which is controlled by parameters in the compartment model. In order to provide model simulations, two approaches are used to demonstrate various possibilities of the non-GIS solution and the spatio-temporal modeling in the GIS environment. As an example, the scenario based on ACSL (Advanced Continuous Simulation Language) represents the standard way of dynamic modeling that is extended by sharing data between the simulation tools and the spatial database. For a more complex spatio-temporal analysis, the solution of a compartment model is implemented in the GIS programming environment. According to the measured data, the simulation results support the idea of desorption from relatively contaminated sediments in the eastern Irish Sea mud patches by pore-water exchange and tide and storm activity that is followed by transport in the dissolved phase, which can subsequently leads to accumulation of plutonium ($^{239/240}Pu$) in other regions.

Keywords: Spatio-temporal modeling; radionuclide deposition; GIS.

1. INTRODUCTION

The mechanisms responsible for the observed patterns and behavior of artificial radionuclides discharged in the Irish Sea over a long time period can be explored by using numerical models shared in the environment of the Geographic Information System (GIS). While the GIS methods can describe and analyze spatial interactions in the selected spatial scale, the temporal models based on differential equations simulate the functioning of environmental processes in the time scale. But, both the systems traditionally operate with different data structures and methods. It makes difficulties to explore both the spatial and temporal phenomena in the framework of a single modeling tool. It can be overcome by linking of spatial and temporal modeling tools in the GIS, which supports a wide range of case oriented tools and development programming libraries [1].

Spatio-temporal modeling of radionuclide transport in the marine environment can be solved with existing GIS capabilities, but an extension is needed for time consuming numerical calculations of dynamic models. In order to develop simulation tools, the approach also requires programming skills focused on GIS development platforms. Thus, for a decade, a more traditional way has been represented by linking two separate systems based on simulation tools for computationally intensive numeric tasks [2], and on GIS for spatio-temporal analysis [3].

Since modeling of radionuclide distribution generally depends upon physical principles, the descriptions of these models are in the form of differential equations that implicitly recognize the continuity of space and continuous variables. In order to implement numerical solutions of these models in digital computers, it is needed to propose methods for discretizing the spatial continuity of both the models and the modeled phenomena. The majority of physically based environmental models include fundamental laws of physics dealing with transport in the water column, exchange of contaminants between dissolved and particulate phases, sediment resuspension, sediment transport and mixing of material within the seabed [4]. Many of the fundamentals laws of physic are expressed as partial differential equations that need to be solved in order to find the values of radionuclide concentrations at any place. For some simplified special forms, the solutions can be found by analytical means. But in many cases, the models must be solved numerically, which offers several different types of numerical solutions. The finite difference numerical solution is based on discretization in time and space into small units. Finally, the stepped algebraic solutions for governing differential equations are calculated for each time and space unit. Thus, the fields of radionuclide concentrations generated by finite element solutions divide the study area into units that are homogenous in ways that enable some terms of the governing equations to be simplified. The discretization in space can be based on regular or irregular grids. The irregular grids enable to split the area of interest into parts that indicate similar properties in terms of radionuclide distribution. While the regular grids offer well-established algorithms to process model calculation, the rigid grid structure does not conform to the variability of environment. In many cases, splitting the areas of interest into larger irregular parts is often caused by inaccessibility of all data for setting initial conditions, boundary conditions and model parameters in regular dense grids. Thus, the models based on irregular grids with an ordinary differential equation for each part are developed to simplify complex processes of radionuclide distribution, and to simplify the model simulation.

2. Distribution of Radionuclides in the Irish Sea

The artificial radionuclides reside in the sub-tidal sediments of the Irish Sea due to authorized discharges from the nuclear reprocessing plant. The quantities of the major nuclides have been released into the Irish Sea since 1952 to 1988 [5]. The radionuclides entered the Irish Sea from Sellafield in low-level liquid wastes discharged through a pipeline which terminated 2.1 km beyond the low-water mark. The quantities and relative proportions of nuclides depended on a number of factors varied throughout the history of the Sellafield plant, such as the nature of the fuel, its burn-up time, the nature and length of storage prior to reprocessing, the method of reprocessing and the nature of effluent treatment plants.

Many research projects have been carried out by staff of the Ministry of Agriculture, Fisheries and Food's (MADD) Directorate of Fisheries Research (DFR) at Lowestoft, in support of MAFF's functions under the Radioactive Substances Act, 1960 [6]. The projects are mostly set up for verifications of the liquid radioactive waste discharges to the aquatic environment, and for radionuclide monitoring in the living environment, in order to ensure that the resulting radiation exposure is within nationally accepted limits. Many studies and reports published until now have been focused on the accumulation of radionuclides by biological materials and the subsequent radiation exposure of the public, besides the transport processes in the Irish Sea. For a few decades, a large amount of data has been collected by the staff of DFR. But, individual data sets were collected with respect to the problems prevalent at the time period, using the corresponding methods. Thus, the results of inconsistent sampling techniques, sampling depths, radioanalytical methods, etc. need to be converted and verified to study the geochemical processes controlling the dispersion of radionuclides [7].

The radionuclides are dispersed in the Irish Sea that is the semi-closed body of water bounded by coast lines in the south and in the west, Figure 1. The St George's Channel in the south and the North Channel are only entrances to the Irish Sea, both indirectly connecting it with the Atlantic Ocean. The deep trough to the west, with a maximum depth of more than 275 m, extends from the St George's Channel to the North Channel with relatively shallow embayments of depth less than 50 m. The estimate of the total volume of the Irish Sea is 2430 km^3. The tide propagates into the Irish Sea from the Atlantic Ocean through both the channels and is able to create vertically homogeneous water column in most areas. In spite of that quantifying the net outflow via channels is complicated, the oceanic inflow via the St George's Channel can be 2.2 km^3 per day and the net outflow through the North Channel ranging from 2 to 8 km^3 per day. Apart from the dominant flow, which is principally north-going, the radionuclide dispersion is affected at the shore by winds and storms. Also, the distribution and origin of major sediment types have been extensively studied and summarized using geophysical and drilling techniques. A number of maps were published by the British Geological Survey and many reports were carried out by Ministry of Agriculture, Fisheries and Food, Figure 2 [8]. The Quaternary history of the Irish Sea mud patches in its various parts is the subject of debates. The results based on radiocarbon dating techniques indicate that recent sediment accumulation has taken place at a few sites, at least, in the eastern Irish Sea mud belt, approximately 1 mm per year over the last 3000-4000 years. But, the sediments are vertically mixed to considerable depth as a result of benthic fauna within the seabed, which is also responsible for the penetration of some radionuclides, such as Pu, Am and Cm to depths of up to more than 1 m in the Irish Sea [9].

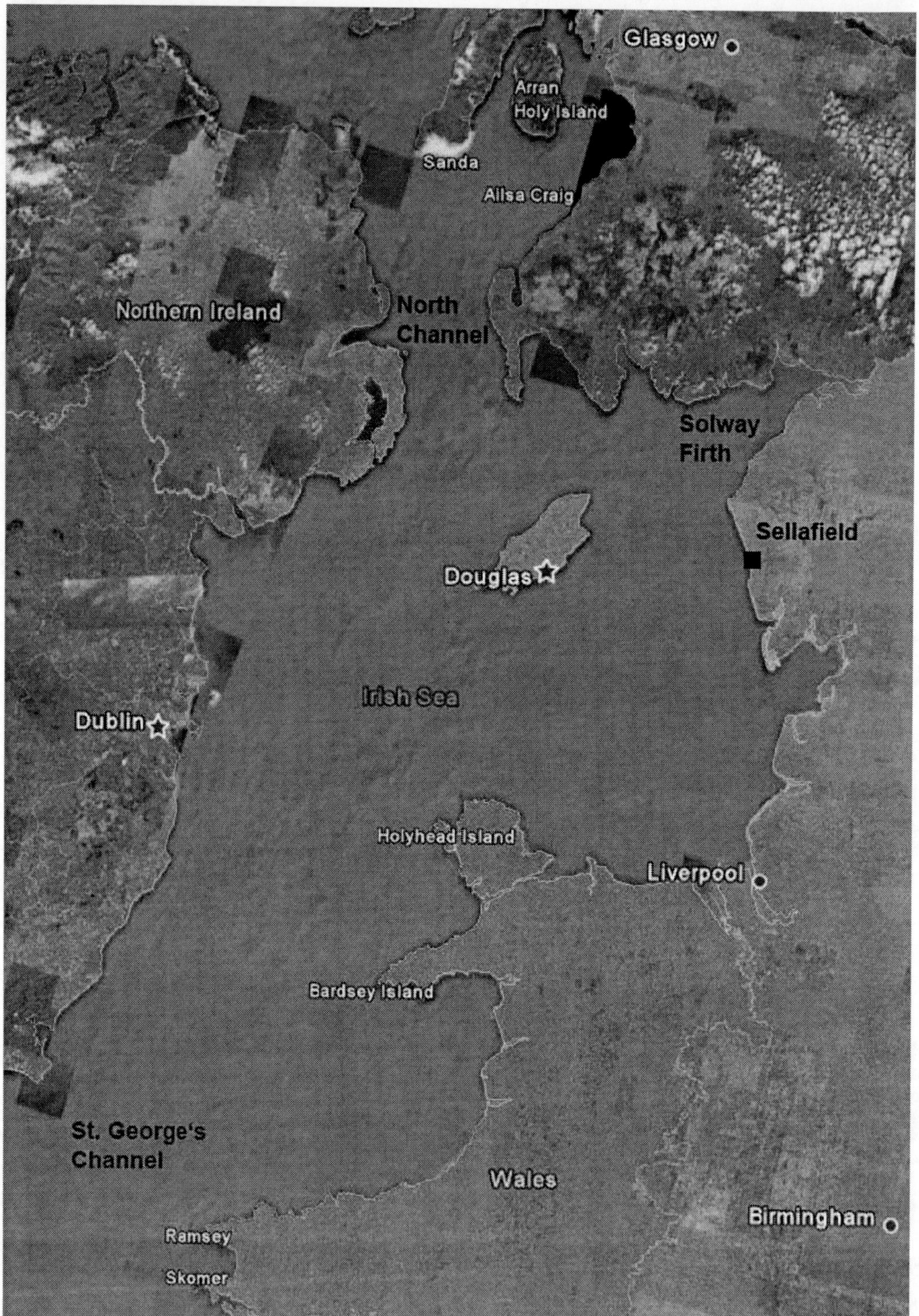

Figure 1. The Irish Sea complemented by the nuclear facilities sited Sellafield and the area of interest Solway Firth in the east part of the coastline (captured from the GoogleEarth™ in 2009).

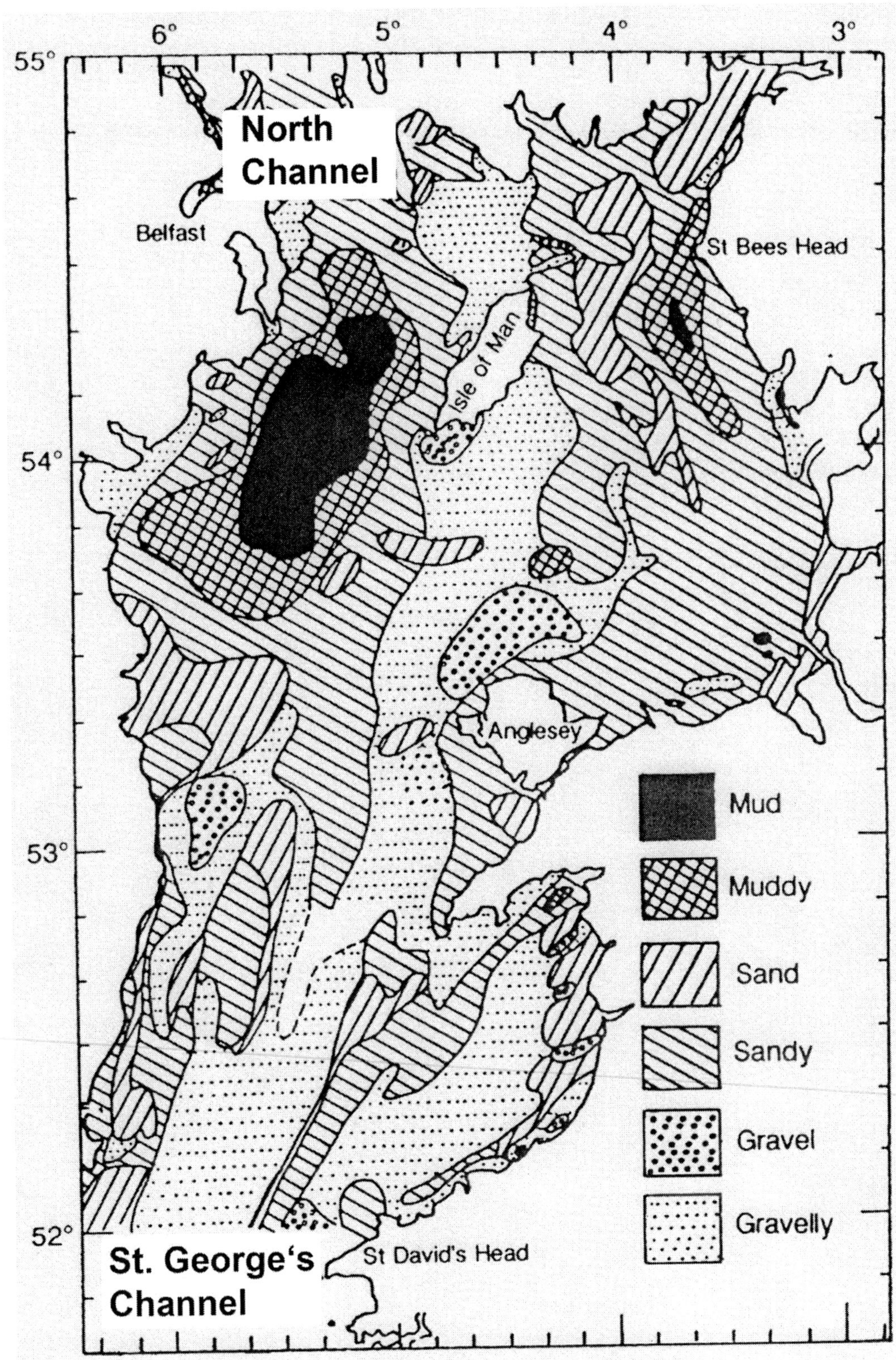

Figure 2. Distribution of fine-graned sediments in the Irish Sea [8].

3. Modeling of Radionuclide Deposition

Modeling of radionuclide deposition in the marine environment is based on a compartment model. Available data for radionuclide transport in the whole area and longtime period have generally necessitated a relatively simple treatment of physical processes in the model. Thus, the Irish Sea was divided into parts that should indicate relatively homogeneous radionuclide concentrations and similar environmental properties in view of geophysical processes. Each subarea is also divided into a few vertical layers: water column, interface and three sediment layers. A schema in Figure 3 illustrates irregular subareas for modeling of transport processes via fluxes between neighbor compartments and vertical layers. The flux estimate between neighbor compartments is based on the difference of radionuclide concentrations multiplied by a transfer coefficient, which can be complemented by other functions in dependence on the environment. Because the fluxes combine many physically distinct transport processes and interactions with the environment, the simple initial representations of radionuclide dynamics is progressively improved in order to give insight into the actual mechanisms responsible for the distribution of radionuclides.

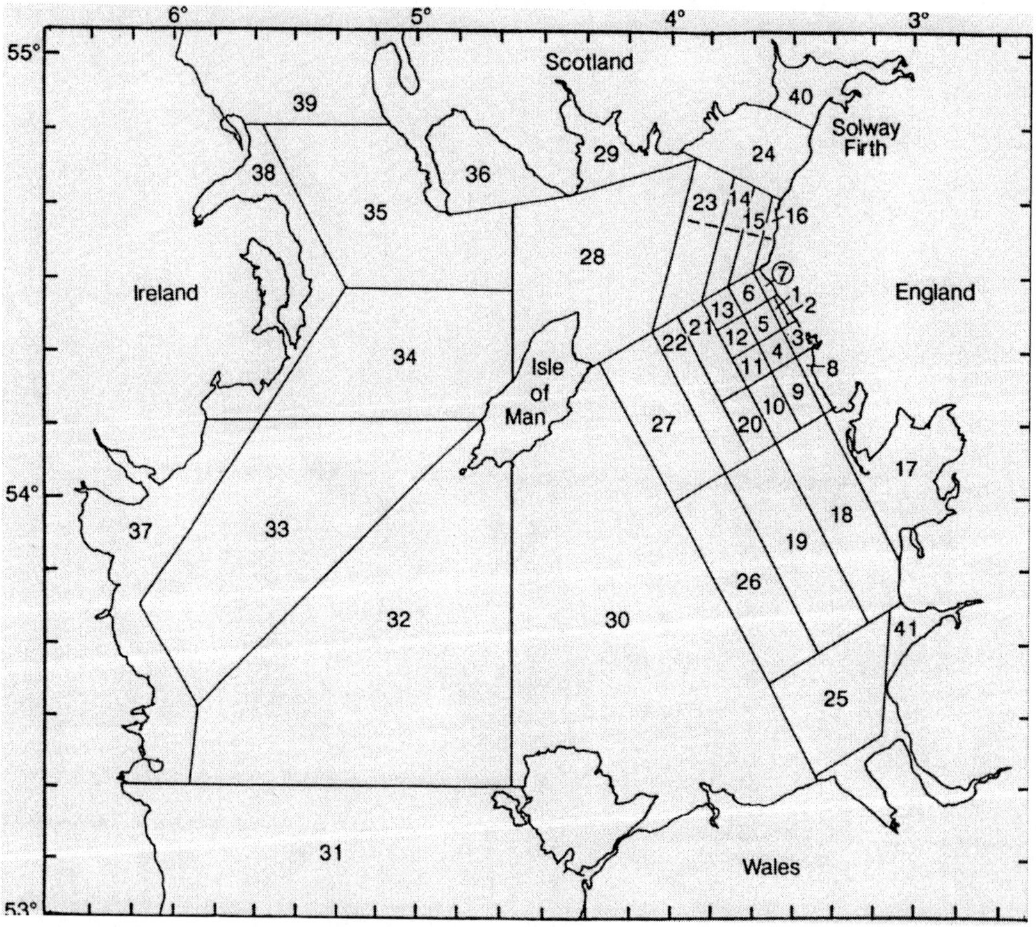

Figure 3. Regional structure of the MAFF MIRMAID compartmental model [8].

In general, transport denotes processes which can determine the distribution of mass, heat or species in our environment. The transport processes are relevant in almost all environmental systems. In the narrower sense, two basic transport processes are represented by advection and diffusion. Advection denotes transport by shifting particles from one place to another by the flow field. Diffusion originates from concentration differences that result in a tendency to equalize concentration levels. In dependence on the environment, there are different diffusion speeds in surface water bodies and in sediments, which is controlled by parameters in the compartment model [11]. In addition to radionuclide transport in the relatively homogenous compartments, the settling and resuspension can move radionuclides from the aquatic environment to the benthic one and vice versa. Both the processes are complicated by a strong interaction with the biotic processes. In order to overcome complex interactions and measurement limitations of radionuclide transport between water bodies and sediments, a simplified mathematical description is given to the mathematical model that can be elaborated in the next steps in dependence on other measurements. It also includes incorporation of time-dependent flow fields and improved treatment of sediment processes. Further steps can attempt to reproduce physical, chemical and biological interactions in a more realistic way, which includes resolving of tidal and wind-driven flows together with wave-current sediment resuspension and transport. But, all these steps need more extensive data collection and observation for precise setting of the model parameters.

In order to demonstrate the radionuclide transport in the Irish Sea, a dynamic model is developed for accumulation and redistribution of the plutonium ($^{239/240}Pu$) over a few decades. In addition to modeling of the radionuclide transport, an implementation of the simplified dynamic model offers a few ways of its solution, such as by a standard way with a continuous simulation language, and by GIS development tools with programming of a standalone software application in the GIS environment. The basic schema of the radionuclide transport among compartments in regions and between layers is in Figure 4.

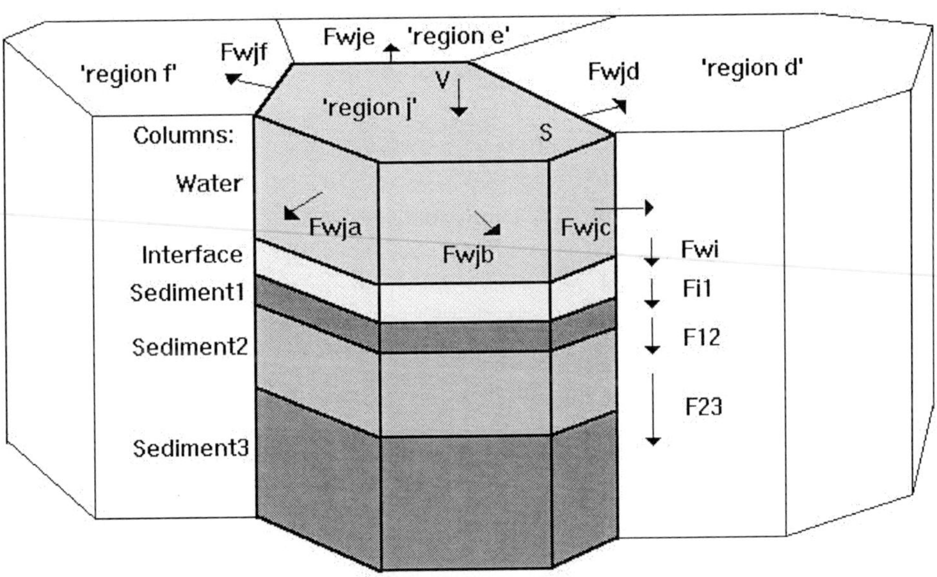

Figure 4. Structure of the compartmental model for modeling of the radionuclide deposition [12].

In case of the radionuclide, the change in the amount depends on: radioactive decay law, and radionuclide rate in and rate out:

$$\frac{dx(t)}{dt} = -kx(t) - RateOut(t) + RateIn(t)$$

where k is a positive constant of proportionality independent on the time t and the amount x. Thus, a basic schema used for the radionuclide accumulation and transport in a region can be described by the ordinary differential equations:

$$\frac{dx_w(t)}{dt} = -kx_w(t) - F_{wi}(t) + \sum\left(F_{wf}(t) + F_{we}(t) + F_{wd}(t) + \cdots\right),$$
$$\frac{dx_i(t)}{dt} = -kx_i(t) + F_{wi}(t) - F_{is1}(t) + \sum\left(F_{if}(t) + F_{ie}(t) + F_{id}(t) + \cdots\right),$$
$$\frac{dx_{s1}(t)}{dt} = -kx_{s1}(t) + F_{is1}(t) - F_{s1s2}(t) + \sum\left(F_{s1f}(t) + F_{s1e}(t) + F_{s1d}(t) + \cdots\right),$$
$$\frac{dx_{s2}(t)}{dt} = -kx_{s2}(t) + F_{s1s2}(t) - F_{s2s3}(t) + \sum\left(F_{s2f}(t) + F_{s2e}(t) + F_{s2d}(t) + \cdots\right),$$
$$\frac{dx_{s3}(t)}{dt} = -kx_{s3}(t) + F_{s2s3}(t) + \sum\left(F_{s3f}(t) + F_{s3e}(t) + F_{s3d}(t) + \cdots\right),$$

where the x variables are amounts of a radionuclide, the k represents radioactive decay, and the F are radionuclide transport fluxes. The amounts of a radionuclide in compartments and the radionuclide fluxes between the compartments are dependent on time t. The amounts of a radionuclide in various phases (depth-averaged dissolved-phase, water-column particulate-phase, layer-averaged bed concentration, pore-water concentration) can be determined by Bq/kg or Bq/m^3 in dependence on available data and measurement techniques. The radionuclide transport fluxes are indicated as the amounts per second. The detailed description of the radionuclide transport fluxes in dependence on amounts of a radionuclide and various abiotic and biotic factors requires more concentrated view and more elaborate measurements focused on particular parts of the area of interest.

In order to provide model simulations, two approaches are used to demonstrate various possibilities of the non-GIS solution and the spatio-temporal modeling in the GIS environment. As an example, the scenario based on ACSL (Advanced Continuous Simulation Language) represents the standard way of dynamic modeling that is extended by sharing data between the simulation tools and the spatial database. The input data are imported from spatial database to the ACSL programming structure. After the simulation, the data outputs are backward imported from binary or text files to the spatial database. In case of the spatio-temporal modeling, the GIS can manage all the data transactions, but a little programming is needed to develop numerical models for solution of the ordinary differential equations.

4. Spatio-Temporal Modeling

ACSL represents one of the simulation languages that are developed for modeling systems described by time-dependent, non-linear differential equations. Their typical application are focused on chemical process dynamics, control system design, plant and animal growth, etc. Working from an equation description of the problem, or a block diagram, the user writes ACSL statements to describe the system under investigation. An important feature of ACSL is its sorting of the continuous model equations, in contrast to general purpose programming languages, such as FORTRAN and C. The ACSL programming code is read by the ACSL translator that translates it to the FORTRAN compile file. Then, the FORTRAN code is compiled, linked with the ACSL runtime library, and executed at an ACSL runtime prompt. At this point, ACSL is waiting for runtime commands that can be entered interactively or by reading a command file. During the simulation, ACSL can generate plots and output data files in dependence on plotting and printing runtime commands. The basic structure of the ACSL programming code is in Figure 5. In the section derivative, the programming code includes statements for setting of the numerical method (ALGORITHM IALG=4, the 4th order Runge-Kutta) and its parameters (MAXTERVAL MAXT=0.01, the step size; NSTEP NSTP=1, the number of steps in a communication interval; CINTERVAL CINT=1.0, data logging rate). The input data, such as the radionuclide discharges, the water flows and the time variable model parameters are readable from tables. Other non-variable model parameters are included into the programming code as constants. The key part is focused on structure of the mathematical model. It contains ordinary differential equations transformed into ACSL statements. There are 41 regions in the area of the Irish Sea. Each region contains a water layer, an interface layer and three sediment layers. Thus, there are 205 compartments and each compartment is represented by one ordinary differential equation in the ACSL numerical model. Finally, the terminate statement (TERMT) defines the program terminating condition, where the independent variable t is the simulation time. The model state variables and derived variables are saved to the ASCII file each communication interval, which is one year (CINTERVAL CINT=1.0), but the simulation step has to be even shorter (MAXTERVAL MAXT=0.01). The ACSL runtime commands start the actual run. They are mainly used to generate the simulation report, to modify saving of the variables, and to manage plotting. After the run, the output data stored in the binary files can be transferred into another computer program in order to provide spatio-temporal analysis, risk assessment and output reports.

Simulation tools, such as ACSL and similar systems, have been developed for a few decades. These packages have been particularly successful for users of personal computers. They provide a set of tools for solving equations with numerical methods, which is a major improvement in comparison to the programming tools, such as FORTRAN, previously used for simulation of environmental models. Nowadays, mathematical software, such as for example MATLAB and its toolboxes [13, 14] offers even more valuable and cost-effective tools that minimize the risk of programming errors by the advanced user environment, and include advanced visualization tools. It allows the user to immediately visualize and animate simulation results. For advanced higher dimensional problems, the MATLAB partial differential equation toolbox or other multi-physics computation tools, such as COMSOL, can be used for solving of a particular model by the Finite Element Method (FEM).

There are other suitable environmental modeling tools for simulation of the radionuclide deposition. For example, STELLA represents a software tool with the icon-based graphical interface that simplifies model building and provides more complex insight into how systems work [15, 16]. After a model creation, model equations are automatically generated and made accessible beneath the model layer. New model features, such as arrays and user built blocks, can simply represent repeated model schemas and multi-level hierarchical model structures.

```
PROGRAM Radionuclide Deposition Modelling 239,240Pu
DYNAMIC
DERIVATIVE
ALGORITHM IALG=4 ! Runge-Kutta 4.order
MAXTERVAL MAXT=0.01
NSTEPS NSTP=1
TABLE
{input of radionuclide discharges}
{model constants}
{differential equations}
{integrations}
END ! of Derivative
{output conversion}
CINTERVAL CINT=1.0 ! data logging rate
CONSTANT tStop=50.0 ! terminate statement
TERMT(t .GE. tStop, 'Stop on time limit')
! output into the data file
{ACSL/ANSI FORTRAN 77 standard code}
END ! of Dynamic
END ! of Program
```

Figure 5. The basic structure of the ACSL programming code.

In addition to simulation languages and mathematical software tools, there are modeling systems for case oriented environmental research, such as groundwater, environmental and water resources modeling. Many of these software tools have long been the choice of government agencies and consulting firms around the world. For example, the Groundwater Modeling System (GMS) provides tools for every phase of a groundwater simulation including site characterization, spatial mapping, model development, calibration, model simulation and final visualization. Modeling tools, such as GMS and others, support both finite-difference and finite-element models in 2D and 3D including MODFLOW, MODPATH, MT3DMS/RT3D etc. The GIS extensions enable to use spatio-temporal objects and associated data to conceptualize model projects, and link data to GIS in order to provide spatio-temporal analysis and visualization extended by GIS data and layers for background maps. In case of radionuclide modeling, similar modeling systems can be developed to solve accumulation and redistribution of plutonium ($^{239/240}$Pu) or other radionuclides.

5. Spatio-Temporal Modeling with GIS

In order to create a more complex spatio-temporal model, the solution of the dynamic model was implemented in the GIS environment. Recent GISs offer programming tools for development of the scripts that can solve simplified numerical models. The basic model schema is in Figure 6. It contains the main data structures and data flows for solution of accumulation and redistribution of plutonium ($^{239/240}$Pu) in the framework of the proposed compartment model. The input data sets are formed by initial values of state variables, and by time dependent model variables. After the model simulation, the time series of state variables are saved in the output table. Each model compartment is linked to an individual row in the output table. The columns represent specified time points chosen from the overall simulation period. Consecutively, the data can be transferred to the time tables for building of charts with time series, or to the area tables for creating map layers, Figure 7.

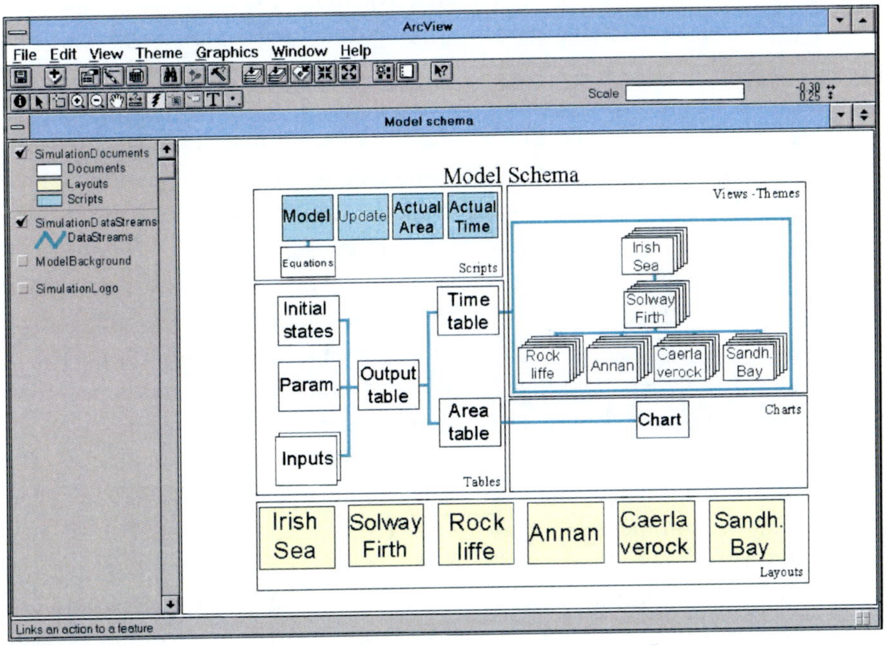

Figure 6. The basic model schema in the GIS-ArcView environment [12].

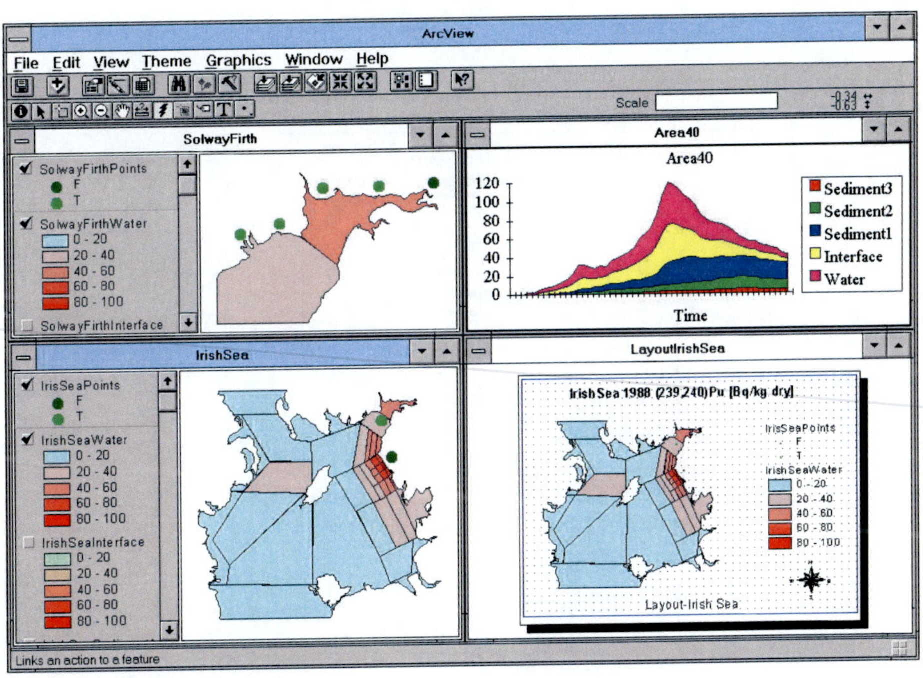

Figure 7. An example of simulation outputs that contain the detail view, the time series, the overall view and the layout view in the GIS-ArcView environment [12].

The basic structure of the script for simulation is in Figure 8. It contains a few steps: initialization, calculation and output data storage. The initialization contains the user interactive input for setting the whole simulation period (TStop), the time step of a numerical method for solution of the ordinary differential equations (TStep), the sampling interval (NStep) used to save state variable values to the output table, the choice of a numerical method (Numerical Method: Euler, Runge-Kutta fixed step second order RK2, or Runge-Kutta fixed step fourth order RK4), and a degree of the Lagrangian interpolation for time dependent input variables [17]. The Runge-Kutta fixed step, fixed order methods are more accurate, and are mostly applied to the final simulations, because they include more time consuming calculation in the GIS's scripts. The first order Euler method is least accurate, but uses less computer time. The next step is the initialization of the state variables (X), temporary state variables (TX), time, model parameters, and other time dependent variables. The temporary state variables are used for saving new state variables. After initialization, state variables and actual time are saved into the output table, and the stop flag is tested. If the actual time value is less then a conditional constant (TStop), the states and time variables are transferred to the temporary variables and the numerical method calculate new state variables for the next step over the time interval (TStep). The internal scripts are called for calculation of the right side of differential equations, and for interpolation of the time dependent variables. Then, the actual time is incremented by (TStep) and the values of all state variables are written each nth step (NStep) out to the output table. Then, the program loops, and the Stop flag is tested again. After the simulation period, a number of scripts are used for data transfer from the output table to the time series (TIME TABLE) and to the layer attributes (AREA TABLE). Thus, the simulation outputs can be displayed in the charts and map layers. Finally, map layouts are generated to provide visualization together with other spatial data.

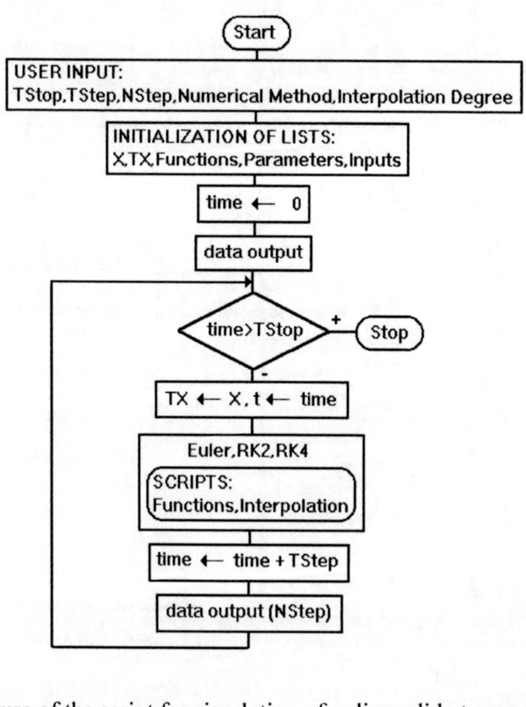

Figure 8. The basic structure of the script for simulation of radionuclide transport [12].

6. CONCLUSION

An exploration of the mechanisms responsible for the behavior of artificial radionuclides, namely the plutonium ($^{239/240}Pu$) in the Irish Sea over a long period was studied using numerical models that were implemented in the various computing environments. In addition to a creation of the mathematical models, the question of the transport mechanisms was addressed using the simulation by ACSL, and the simulation by a numerical model implemented in the GIS environment. Both the ways can provide a simplified model solution of accumulation and redistribution of plutonium ($^{239/240}Pu$) in the framework of the proposed compartment model, but the GIS environment offers more complex processing of spatio-temporal data. It offers sharing of the simulation data with other risk assessment methods included in the monitoring of water and terrestrial ecosystems, and public health research.

In order to compare modeling results with other studies and run simulation in the GIS environment more efficiently, the simplified compartment model was created to solve accumulation and redistribution plutonium ($^{239/240}Pu$) in a few layers divided into irregular areas. The processes for seabed accumulation was based on scavenging of the dissolved phase from the water column by tide and storm re-suspension of flocculated mud and sand, and on interaction with pore water that is undergoing exchange with the upper water column. Thus, according to the measured data, the simulation results support the idea of desorption from relatively contaminated sediments in the eastern Irish Sea mud patches by pore-water exchange and tide and storm activity that is followed by transport in the dissolved phase, which can subsequently leads to accumulation of plutonium ($^{239/240}Pu$) in other regions in the Irish Sea [4]. Because the mechanisms of the behavior and movement of radionuclides in the Irish Sea are rather complex, there is still considerable uncertainty in the framework of the model validation. This suggests the need for further monitoring and the more precise model calibration in order to explore in a more detail the mechanisms of radionuclides distribution.

The model solution by simulation languages offers efficient numerical calculations and no needs for programming of numerical methods. But, in spite of that the simulation language, such as ACSL, has many extensions for control theory analysis, there is limited support for data exchange in the framework of sharing information with other information systems and spatial databases.

In case of using the spatio-temporal modeling in the GIS environment, this platform offers better support for more complex decision-making processes and is easy to use and understand. Generally, the GIS is able to access and manage large volumes of heterogeneous spatio-temporal data. But, there is still limited support for time consuming calculations, such as the solution ordinary differential equations, in the programming libraries. Thus, there is need to develop extensions for numerical modeling in the GIS environment. Recent GISs offer using of higher programming languages and object oriented libraries. But, in case of using scripts, the numerical models have to be simplified in order to run simulation in a more efficient way, and to manage middle-size data sets.

Simulation of the radionuclide deposition has been tested in the ArcView 3.x environment, which is the predecessor of ESRI's ArcGIS, by building the case oriented scripts for numerical simulation. The reference simulation system was represented by simulation programs written with simulation language ACSL. In spite of that the simulation in the ACSL environment used less computer time and enabled to choose more numerical algorithms, the simulation in the GIS environment provided a more complex support for the purpose of spatio-temporal radionuclide modeling that offered methods of the exploratory spatial data analysis, creation of the spatial interpolations, methods of advanced visualization and creation of the final map layouts.

In this case, spatio-temporal modeling managed by GIS can support environmental planning and policy making [18, 19, 20]. GIS by itself can provide direct access to a large volume of data. Its functionality includes a wide range of analytical methods that can be used for interpretation, scenario analysis and forecasts. Thus, it can bring the best available knowledge to bear on decision making processes and reach a broad audience.

ACKNOWLEDGMENT

The presented research was carried out in the framework of the project Environmental Modeling with GIS supported by The Royal Society in London. The spatio-temporal data are based on reports carried out by the Ministry of Agriculture, Fisheries and Food's (MADD) Directorate of Fisheries Research (DFR) at Lowestoft. Spatio-temporal data were processed by ACSL and ESRI's ArcView at University of Glasgow. Initial research results focused on modeling of radionuclide deposition was presented at the 1st International Conference on GeoComputation at University of Leeds in 1996. Initial research results dealing with spatio-temporal modeling in the GIS environment was demonstrated at the 11th ESRI European User Conference in London in 1996. Some actual research results focused on mechanisms of radionuclide deposition in the Irish Sea originated from papers published by the Centre for Environment, Fisheries & Aquaculture Science, Suffolk, United Kingdom.

REFERENCES

[1] Goodchild, M.F., Steyaert, L.T. and Parks, B.O. (1996). *GIS and Environmental Modeling: Progress and Research Issues*. Fort Collins: GIS World, Inc.

[2] Holzbecher, E. (2007). *Environmental Modeling Using MATLAB*. Berlin: Springer.

[3] Maguire, D.J., Batty, M. and Goodchild, M.F. (2005). *GIS, Spatial Analysis and Modeling*. Redlands, California: ESRI Press.

[4] Aldridge, J.N., Kershaw, P., Brown, J., McCubbin, D., Leonard, K.S. and Young, E.F. (2003). Transport of plutonium ($^{239/240}$Pu) and caesium (^{137}Cs) in the Irish Sea: comparison between observations and results from sediment and contamination transport modeling. *Continental Shelf Research 23*, 869-899.

[5] McCartney, M., Kershaw, P.J., Woodhead, D.S. and Denoon, D.C. (1994). Artificial radionuclides in the surface sediments of the Irish Sea, 1968-1988. *The Science of the Total Environment 141*, 103-138.

[6] Great Britain-Parliament (1960). Radionuclide Substances Act. London: HMSO, 28 pp.

[7] Lucey, J.A., Gouzy, A., Boust, D., Vintró, L.L., Bowden, L., Finegan, P.P., Kershaw, P.J. and Mitchell, P.I. (2004). Geochemical fractionation of plutonium in anoxic Irish Sea sediments using an optimised sequential extraction protocol. *Applied Radiation and Isotopes 60*, 379-385.

[8] Kershaw, P.J., Pentreath, R.J., Woodhead, D.S., Hunt, G.J. (1992). A review of radioactivity in the Irish Sea. Suffolk: Ministry of Agriculture, Fisheries and Food.

[9] Kershaw, P.J., Swift, D.J., Pentreath, R.J., Lovett, M.B. (1984). The incorporation of plutonium, americium and curium into Irish Sea seabad by biological activity. *Science of the Total Environment 40*, 61-81.

[10] Periáñez, R. (1999). Three-dimensional modelling of the tidal dispersion of non-conservative radionuclides in the marine environment. Application to 239, 240Pu dispersion in the Irish Sea. *Journal of Marine Systems 22*, 37-51.

[11] Clark, M.M. (1996). *Transport Modeling for Environmental Engineers and Scientists*. New York: John Wiley & Sons.

[12] Matejicek, L. (1996). Modelling radionuclide deposition in the Irish Sea. London: 11th ESRI European conference.

[13] Littlefield, B.L. and Hanselman, D.C. (2004). *Mastering Matlab 7*. USA: Prentice-Hall.

[14] Dabney, J.B. and Harman, T.L. (2003). *Mastering Simulink*. USA: Prentice-Hall.

[15] Deaton, M.L. and Winebrake, J.I. (1999) *Modeling of Environmental Systems*. New York: Springer.

[16] Ford, A. (1999). *Modeling the Environment: An Introduction To System Dynamics Modeling Of Environmental Systems*. Washington: Island Press.

[17] Press, W.H., Teukolsky, S.A., Vetterling, W.T. and Flannery, B.P. (1992). *Numerical Recipes in C: the Art of Scientific Computing*. Cambridge: Cambridge University Press, 2nd edition.

[18] Cromley, E.K. and McLafferty S.L. (2002). *GIS and Public Health*. New York: The Guilford Press.

[19] Arctur, D., Zeiler, M. (2004). *Designing Geodatabases: Case Studies in GIS Data Modeling*. Redlands, California: ESRI Press.

[20] MacDonald, A. (2001). *Building a Geodatabase*. Redlands, California: ESRI Press.

In: Environmental Modeling with GIS
Editor: Lubos Matejicek, pp. 17-32

ISBN: 978-1-60876-363-4

Chapter 2

DYNAMIC MODELING OF SURFACE WATER POLLUTION CAUSED BY THE INDUSTRIAL POLLUTANT RELEASE

Lubos Matejicek
Institute for Environmental Studies, Charles University in Prague, Faculty of Natural Science, Benatska 2, 128 01, Prague, Czech Republic
Lenka Tlapakova
Institute for Environmental Studies, Charles University in Prague, Faculty of Natural Science, Benatska 2, 128 01, Prague, Czech Republic

ABSTRACT

Dynamic modeling focused on surface water pollution by accidental release is considered in the context of the GIS tools. Integration of dynamic modeling in the GIS environment is carried out by the AVSim, the software tool originally developed in the ArcView with the AVENUE programming language, and actually implemented in the ArcGIS. The presented compartment model demonstrates the AVSim capabilities to solve a simplified one-dimensional model that uses a one-dimensional representation of the river. It assumes a prevailing one-directional flow that causes varying properties of the water body along its direction. The varying properties of the water body along its flow direction are approximated by a chain of stocks of the pollutant that represent connected sections of the river. Before creation and validation of the compartment model, the linear interpolation is used to explore time changes of pollutant concentrations in a few river sections neighboring downstream the accidental release. Finally, two simulation studies are presented. The first study is based on no downstream contamination removal after the accidental release. The second study includes the pollutant removal by the remediation procedure, but only the main processes are included into the simplified compartment model. Finally, the results and methods of modeling are discussed.

Keywords: Spatio-temporal modeling; surface water pollution; GIS.

1. Introduction

For a few decades, models focused on surface water pollution have been used to predict pollutant changes with time and at different points downstream of the release pollution point. Time and distance downstream represent the independent variables. The amounts of pollutant or concentrations are the dependent variables. In order to account for variation in the horizontal and vertical dimension, the models must provide an acceptable division of the space to simulate physical processes of the pollutant dispersion. The simplest geometric representation is the zero-dimensional model that can simulate the system represented by a point and the only possibility given to the system is time change of the pollutant amount. It is very often used as a first approximation of the system behavior, and to simulate water pollution of shallow and small water bodies where stratification does not occur and horizontal homogeneity is assumed. For a heterogeneous water network, one-dimensional models use a one-dimensional representation of the system. It is assumed a prevailing one-directional flow that causes varying properties of the water body along its direction. Systems based on streams or rivers are mostly simulated by a one-dimensional model. In case of larger systems with sensible variation of the properties, vertical and horizontal division is required to provide two or three-dimensional representation.

On a temporal basis, water pollution models can be distinguished as being either in a steady state or in dynamic one. If the variables and forcing functions of the system do not change in time, they can only show a variation in space distribution of the properties considered with the distributed parameter model, otherwise it is simulated as a zero-dimensional system.

The procedure for selecting the appropriate time and space scale depends on spatial and temporal variability. It demands an understanding of the dominant processes occurring within the system. Thus, the final mathematical description requires a good definition of the conceptual model including the appropriate time and space scale.

Modeling of water pollution by the transport of mass in fluids is one of the most relevant physical processes of environmental systems. Thus, it is important to know how a pollutant moves within a medium, how it leaves a liquid phase to reach a gas and/or solid phase, and which concentration it can assume in a given place and time. The major processes of mass transport include some combination of advection, diffusion and dispersion [1, 2, 3].

Advection represents one of the ways for movement of a pollutant in water environment by the substance moving solidly with the fluid in one direction without varying its concentration. The theoretical laminar flow in a pipe can be observed approximately in the motion of the water in a large river flowing slowly in a rectilinear branch of its bed. But in case of water flowing in a mountain creek, it cannot be described by advection because mixing of the pollutant and water by a large turbulence is the dominant process of the transport.

Diffusion is represented by the movement of a pollutant due to Brownian motion of water molecules causing the random motion of the pollutant molecules. It has a tendency to minimize gradients of pollutant concentration in water moving the pollutant from a region of high to low concentration. Finally, dispersion is the combination of the described main processes of the mass transport that form advection and diffusion. It is the process in a majority of surface water pollution modified by the real conditions.

2. Dynamic Modeling in the GIS Environment

In order to explore pollutant transport, a simplified one-dimensional compartment dynamic model is developed for the propagation of pollutants through the selected water stream segments in the basin of the river Zelivka in the Czech Republic. For risk assessment and decision-making processes, the compartment model originated from the RIVER TOXINS model [4] is linked to the Geographic Information System (GIS), which offers to use a wide range of spatio-temporal analyses.

In general, environmental modeling mostly deals with space and time, because describes principles of the mass and energy motion. Such processes require data about the environment and modeling results provide additional data to enrich environmental description. Thus, GISs become a suitable environment for spatio-temporal modeling. In order to provide spatio-temporal modeling, there are a few approaches to integrate spatially distributed environmental models and GIS. They range from linkage through shared data files to building dynamic models as complex analytical modules into the fully functional GIS. In case of sharing data files, two separate systems, the GIS and the dynamic model, just exchange files. Basically, the model obtains some of its input data from the GIS and produces some of its output data for import and further processing and display with the GIS. This approach requires little if any software modification, but the significant amount of manual tasks is needed to provide the data exchange. The next step of integration provides a common interface, information sharing and transfer between modeling tools and GIS. It can be achieved by higher–level application languages built into the GIS. Recent object-oriented programming tools enable to develop more efficient structures that can describe real objects in a more realistic way. Thus, the rivers, pollution sources and measurement control points are described by corresponding objects that include specific data structures and methods. In case of ESRI's ArcGIS, the programming library ArcObjects supports a wide range of spatial objects that can be extended by temporal objects for dynamic modeling [5].

The dynamic model described in this paper is solved with the AVSim modeling tools (ArcView Simulation) [6, 7, 8]. It is an extension of the desktop GIS ArcView for calculation of the compartment spatially distributed models by ordinary differential equations. Integration of the modeling tools in the GIS environment enables central data management for dynamic modeling, spatial processing, data series analysis and visualization in the digital map layers in a more efficient way.

As an example, a dynamic model has been implemented in the AVSim. It is focused on the accidental release of about 1-5 tons of the brown coal generator tar into the waste treatment plant, and consecutively, into the river Bela and Zelivka. After the accidental release (Pelhrimov, 29 December 1998), the remediate procedures were started. It also included monitoring of the chemical compounds at the selected profiles of the rivers Bela and Zelivka, Figure 1. Mainly, the phenol concentrations were monitored 7 days after the accident. In the next period, the concentrations of chemical compounds indicated the permanent decrease and the time table of monitoring events was changed. The first period of 7 days was used for spatio-temporal modeling in order to provide the model setting and validation. The dynamic compartment model was developed for prediction of the pollutant transport. The model output indicated the concentration downstream changes in the first period of 7 days and the pollutant decrease in the next days.

Figure 1. The aerial image of the area of interest complemented by the contamination source (Pelhrimov) and control points. The remediate procedures and monitoring started promptly, because the water supply reservoir Svihov in the basin Zelivka represents the most important source supplying the capital Prague in the Czech Republic.

3. Modeling of the Pollutant Transport

It is assumed a prevailing one-directional flow. The varying properties of the water body along its flow direction are approximated by a chain of stocks of the pollutant that represent connected sections of the rivers, Figure 2. Each stock is connected to the next one with a transfer variable F that is controlled by a dwell time T, which represents how long a particles of pollutant and of river water stay in that particular section. The stocks represent the amount of pollution X at any time.

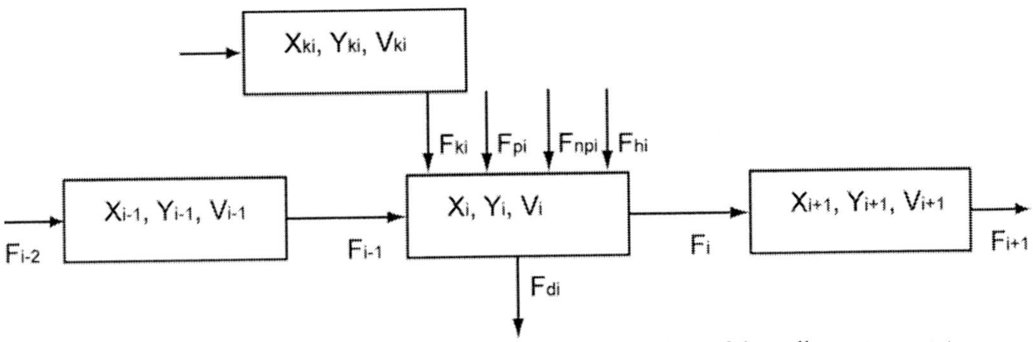

Figure 2. Basic schema of a compartment chain for dynamic modeling of the pollutant transport.

In order to provide mathematical description, the dynamic model is divided into compartments. Each compartment describes processes as a function of inputs and outputs without attempting to determine the precise spatial distribution of the processes within the section [9]. The state variable is the volume of pollutant X [m^3]. Likewise in the real system, each compartment is connected to the next one with a controlled transfer variable F [m^3/s]. State variables X and transfer variables F are in cubic meters of the pollutant in order to standardize the physical units in the mathematical model. Thus, the measured pollutant concentrations, originally denoted in milligrams per liter, must be transformed to the appropriate units. The transfer variable F depends on the water flow Q [m^3/s] and the water volume in the section V [m^3]. The dwell time can be estimated as the river water volume V in the section divided by the water flow Q. The similar estimation was used in the model RIVER TOXINS solved by the simulation tools Stella [4], but the compartment model described in this study is focused on far more sections that are represented by compartments in the dynamic model. In addition to a main source of pollution, the approximate estimates of other potential point and non-point pollutant sources are included together with the river inflows. The main chain of the sections starts at the point of accidental release, and is directed downstream. The corresponding compartments in the main chain contain variables marked as X_i, Y_i, V_i. The variables included in the compartments in the river inflows are marked as X_{ki}, Y_{ki}, V_{ki}. The estimates of the potential point sources of pollution are marked by F_p and the non-point sources by F_{np}. The variable F_h is considered as a potential direct input of the pollutant in each compartment, but in case of the described model, only the first compartment contains the direct input of the pollutant, the accidental release. Chemical or biological transformation and accumulation to the food chains are represented by the outflow variable F_d in each compartment. The compartment model demonstrated in this paper represents an alternative to other studies, which deal with contamination of the basin Zelivka and explore various other potential accidents in the basin used for the water supply. Many of the studies were carried out by the T.G. Masaryk Research Institute in Prague in the Czech Republic.

The compartment model is described by the ordinary differential equations:

$$\frac{dX_i}{dt} = F_{i-1} - F_i - F_{di} + \sum F_{ki} + \sum F_{pi} + \sum F_{npi} + \sum F_{hi}$$

where X_i [m^3] is the volume of the pollutant in the ith section, F_i [m^3/s] is the transfer variable, F_{di} [m^3/s] is the pollutant outflow, which represents chemical and biological transformation and accumulation, F_{ki} [m^3/s] is the pollutant input from the kth river inflow, F_{pi} [m^3/s] represents the pollutant input from the point source of pollution, F_{npi} [m^3/s] is the estimate of the pollutant input from the non-point source of pollution, finally, F_{hi} [m^3/s] is the direct pollutant input to the basin, such as at the site Pelhrimov in the following case study. The transfer variable F_i [m^3/s], which represents the pollutant transport from the ith section to the $i+1^{th}$ section, can be estimated by the following expression:

$$F_i = X_i T_i = X_i \frac{Q_i}{V_i}$$

where T_i [s] is a dwell time, Q_i [m^3/s] is the water flow at the output of the ith section and V_i [m^3] is the water volume of the ith section. Finally, the pollutant concentration in the ith section is estimated as:

$$Y_i = \frac{X_i}{V_i}$$

where X_i [m^3] is the volume of the pollutant and V_i [m^3] is the volume of water in the ith section. The formula also includes the unit recalculation to the milligrams per liter or another appropriate physical unit compatible with the pollutant monitoring.

4. Simulation of the Dynamic Model in GIS

In order to provide simulation, the compartment model has to be transferred into the numerical model that is implemented in the GIS environment. Originally, the numerical model presented in the following case study was based on the AVENUE programming tools in the ArcView environment. Nowadays, it is possible to use ESRI's ArcObjects in the ArcGIS environment. Thus, the higher programming languages and the ArcObjects library can support numerical calculations in a more efficient way.

The GIS can use spatio-temporal data in the framework of map layers, attribute tables, charts, map layouts and user oriented scripts, Figure 3. The views include spatial data in the map layers focused on the river basin and the pollutant transport. In order to provide numerical modeling, the GIS project can manage the map layers of river sections, point and non-point sources of pollution, and control points for pollutant measurements. To provide a more complex spatio-temporal analysis, other map layers can be included into the GIS project, such as aerial photos, satellite images, existing digital thematic map layers, digital terrain models, map layers based on surface analysis (spatial interpolation, surface flow direction and accumulation, map algebra for multi-criteria analysis).

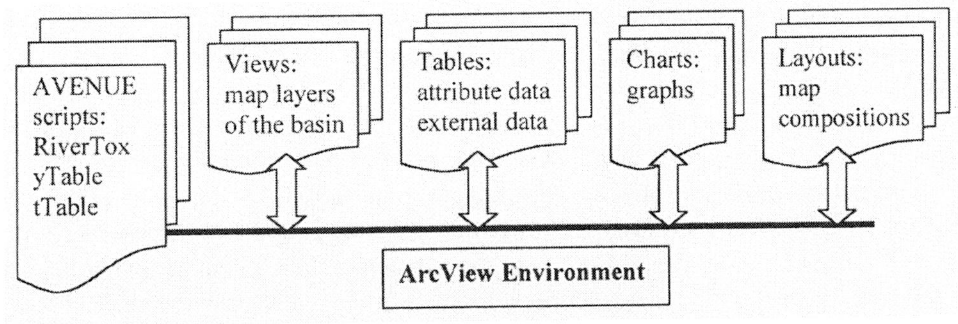

Figure 3. Spatio-temporal data structures in the GIS environment [8].

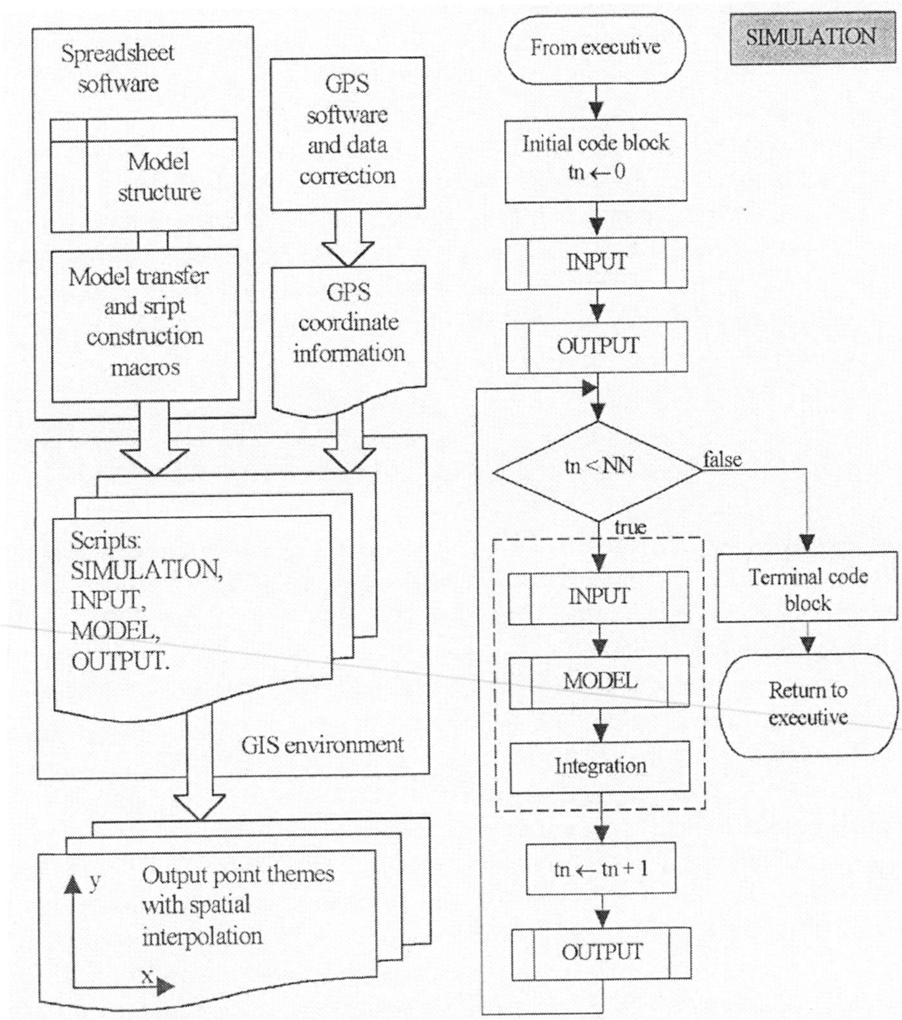

Figure 4. Basic structure of the compartmental model for modeling of the pollutant transport [8].

The numerical solution of a mathematical model is based on AVENUE scripts or programming structures supported by ESRI's ArcObjects. The output data are organized in the attribute tables for display in the map layers, and spatio-temporal analysis (yTable- time series of the pollutant concentrations are organized into the column for each control points; tTable- time series of the pollutant concentration is organized in the row for each control point). The basic structure of a numerical simulation is in Figure 4. The compartment model inputs are formed by differential equations and initial conditions. The spatial data are based on existing digital map layers and spatial features, such as pollution sources and control points that can be located by the GPS. In case of the presented AVSim modeling tools, the compartment model can be also written outside of the script and then transferred into the script by a numerical model builder. After the simulation in the GIS environment, the output map layers are generated in order to display transport of the pollutant in the basin. The script for simulation of the dynamic model is started in the GIS environment. After the initial code block for setting of the simulation parameters (maximum of simulation steps, step interval and numerical method) the input section defines the initial state value and the output sections saves the initial values into output data. Then, the actual number of simulation steps *tn* is compared with the maximum number of simulation steps *NN*. If the flag condition is true, the integration routines provide calculation over a simulation step by using the actual input variables and the model part. The integration routines return the state variables advanced through the simulation step. After the increment of the simulation step counter and saving the actual state variables into output data, the calculation is repeated and condition flag is re-executed. If the flag condition is false, the program control transfers to the terminal code block. On passing out this section, the control is returned to the GIS executive. Then, the further scripts or functions can be executed in order to convert output values, to provide spatio-temporal analysis, and to visualize the output data.

5. Data Processing and Exploratory Analysis

Before creation and validation of the compartment model, the linear interpolation is used to explore time changes of pollutant concentrations in a few river sections neighboring downstream the accidental release, Figure 5-7. Promptly, after the pollutant release, the remediation and monitoring processes were started to reduce the contamination effects. The section 91 is the wastewater tank at the treatment plant. Thus, after the pollutant release, the concentrations are increasing for the period of 7 days. The high concentration in the downstream section 27 in the 1st day after the accident is caused by the immediate partial pollutant release, before the remediation started to takes effects. In spite of that the remediation was started, some pollutant residue was measured lately in the section 27. The pollutant concentrations in the section 17 indicate contamination by the partial pollutant release into the section 27 in the first period. Then, the concentrations are increasing with the peak in the 9th day, which is caused by the transport and accumulation processes from the section 27. After a period of 16 days, monitoring procedures were reduced, because the levels of pollutant concentrations were decreased. This exploratory analysis is used for the model proposal, and for initial setting of the state variables and model parameters. Unexpected high levels of pollutant concentrations in the section 17 indicate potential measurement defects.

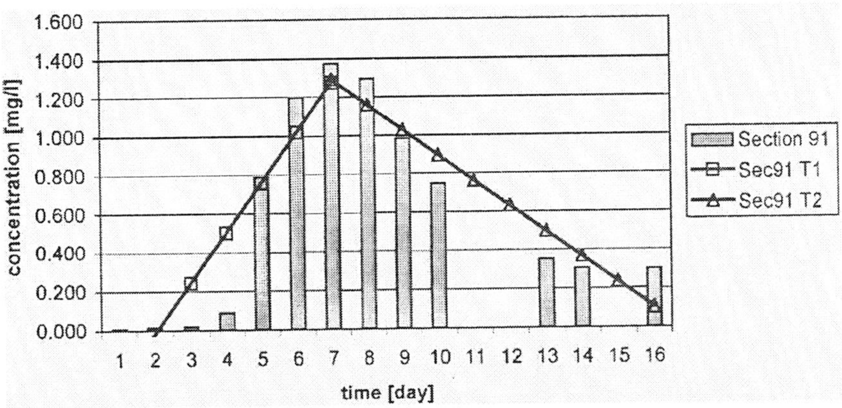

Figure 5. The time series of pollutant concentrations in the section neighboring the release accident, it is divided into two time periods with the increasing and decreasing trends [8].

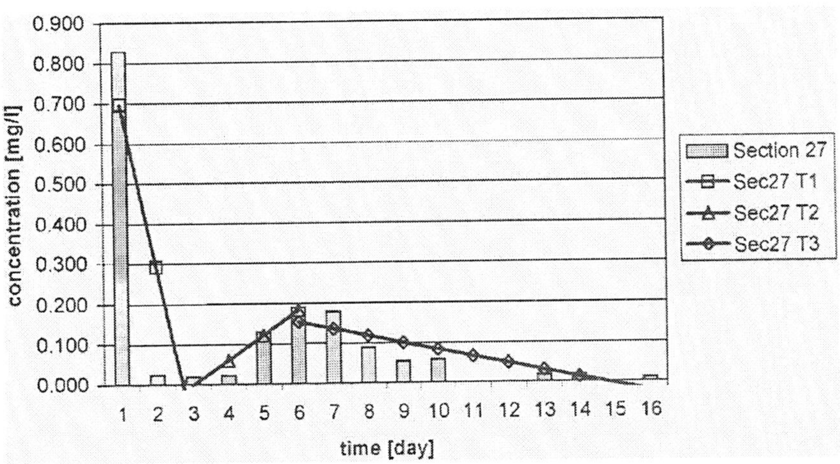

Figure 6. The time series of pollutant concentrations in the next section, it is divided into three time periods with the increasing and decreasing trends [8].

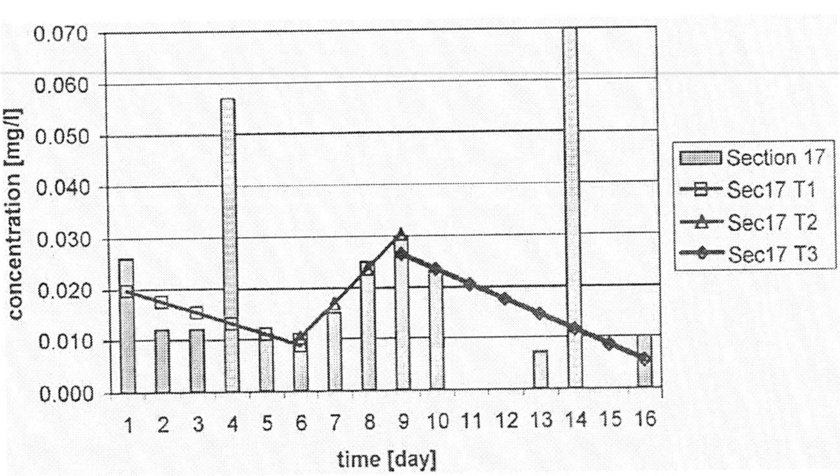

Figure 7. The time series of pollutant concentrations in the next section, it is divided into three time periods with the increasing and decreasing trends [8].

6. Spatio-Temporal Simulation with GIS

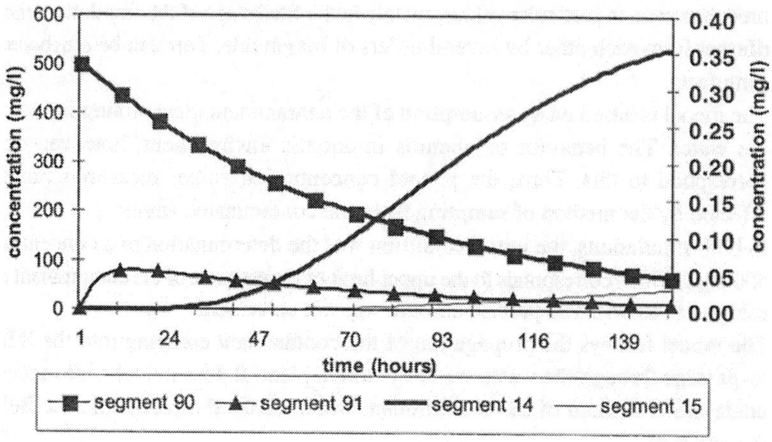

Figure 8. Simulation of the pollutant propagation based on no downstream contamination removal after an accidental release. The left vertical axis indicates pollutant levels in the segment 90 and the right vertical axis indicates pollutant levels in the segment 91 and the far downstream segments 14, 15 [10].

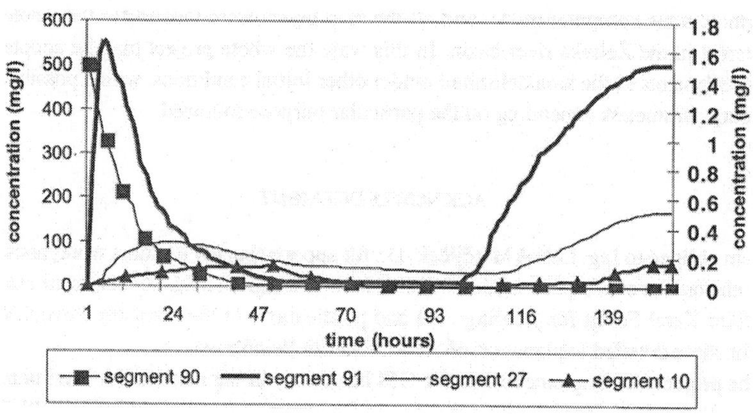

Figure 9. Simulation of the pollutant propagation based on downstream contamination removal by the remediation procedure after the accidental release. The left vertical axis indicates pollutant levels in the segment 90 and the right vertical axis indicates the pollutant levels in the segments 91, 14, 15 [10].

A few simulation studies were carried out by the described compartment model in the GIS environment for an accidental release of the pollutant in the basin Zelivka [10]. Finally, two studies are presented with output time series illustrated in Figure 8-9 and in Figure 10-11. The first simulation study is based on no downstream contamination removal after the accidental release. The time series represented by the curves have identical shapes with the stepwise decreased maximum. The section 90 (segment 90) corresponds to a part of the contamination inlet. An estimate of the initial extreme level of pollutant concentration in the section 90 is based on the amount of a pollutant release and the volume of a waste treatment plant. The pollutant is transported through the basin and accumulated in each section of the water ecosystems. Due to the pollutant accumulation into the sediments and other parts of the water ecosystems, the amounts of pollutant indicate decreasing downstream trends. The second simulation study includes the pollutant removal by the remediation procedure. Thus, only the main processes are included into the simplified compartment model.

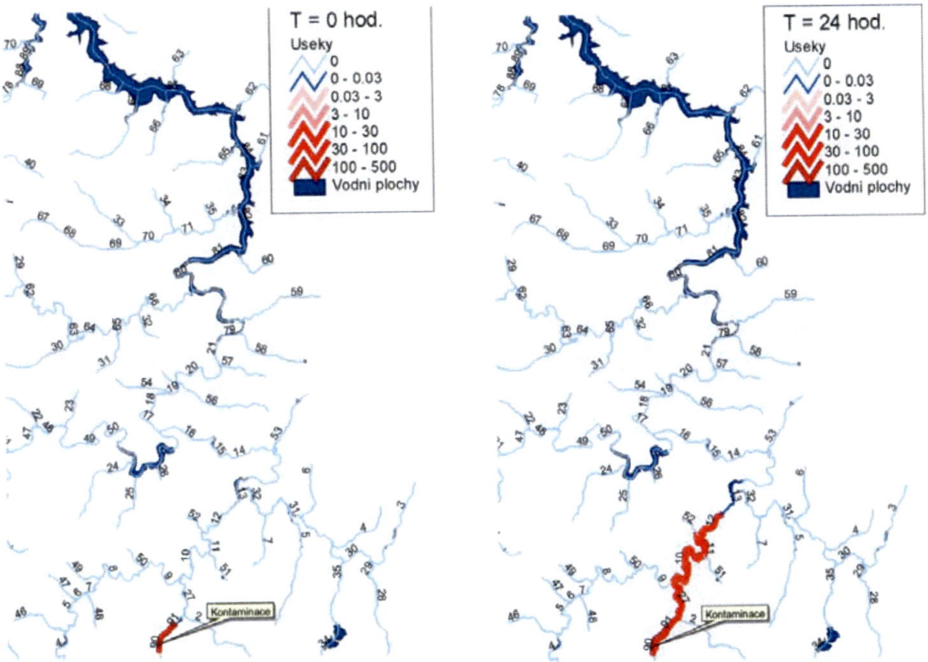

Figure 10. AVSim: pollutant propagation in the basin after the pollutant release T = 0 and T = 24 hours (each section is labeled by its number, the legend is based on the original Czech version) [8].

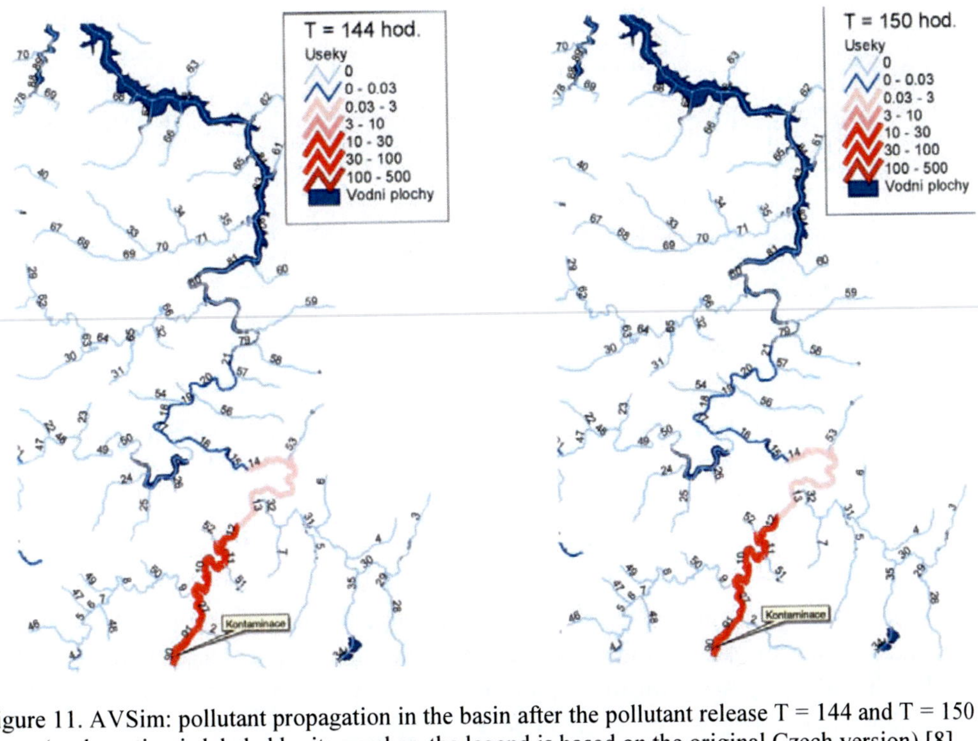

Figure 11. AVSim: pollutant propagation in the basin after the pollutant release T = 144 and T = 150 hours (each section is labeled by its number, the legend is based on the original Czech version) [8].

7. Conclusion

In order to solve the numerical model in the GIS environment, the presented compartment model represents a simplified description of pollutant propagation in the basin. Like in the RIVER TOXINS model, the river is divided into relatively homogeneous sections characterized by the approximate estimates of the water volume and the water flow. Thus, the model is based on an assumption that the pollutant is ideally mixed in each part of the water body. However, the behavior of the real pollutant, such as the brown coal generator tar with phenol pollutants, in aquatic environment does not correspond to these conditions. Also, the pollutant concentrations, mainly phenols, can be affected by the sampling method. The initial concentration of the pollutant in the waste treatment plant was derived from the upper limit estimate of the released pollutant, which was approximately 5 tons (the bottom limit estimate was approximately 1 ton). Due to the lack of experimental data from monitoring systems, the setting of a number of model parameters had to be estimated. All these factors limited a more precise modeling of the pollutant propagation and implementation of more complex models based on two or three-dimensional representation.

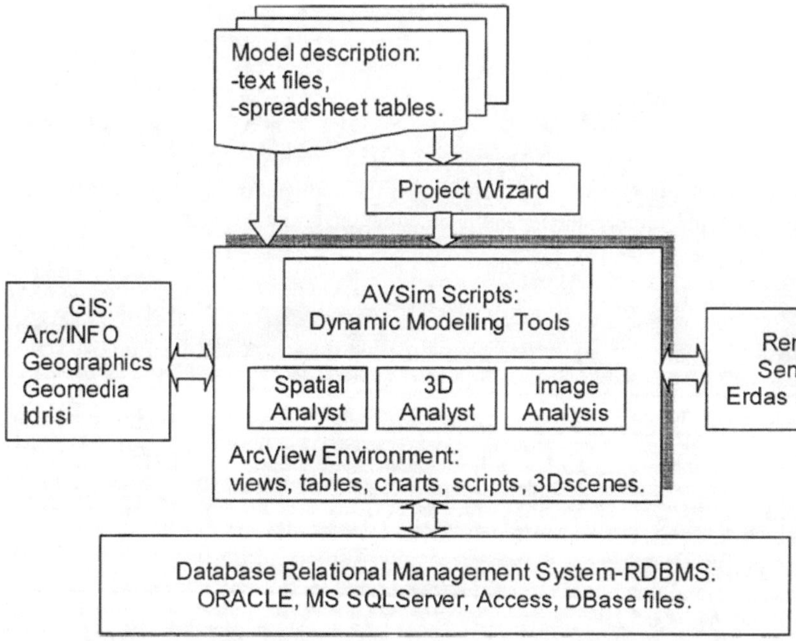

Figure 12. Implementation of the AVSim modeling tool in the GIS. The GIS environment offers sharing spatio-temporal data between other GISs, software tools focused on remote sensing, and database systems that can manage large data volumes in a wide range of data formats [8].

The spatio-temporal modeling in the GIS demonstrates the possibilities to develop more complex tools that can advance scientific research of environmental problems. But, the development of dynamic models needs a more efficient support of the recent GISs. The compartment model for the presented simulation studies was implemented in the ArcView GIS, which is primarily designed to support GIS data display and visualization. Its former programming environment based on AVENUE language was able to provide simplified numerical calculation, but the recent development tools based on ESRI's ArcObjects and higher programming languages are able to solve time consuming numerical algorithms. Actual implementation of the AVSim modeling tools in the GIS is in Figure 12.

The modeling tool AVSim is planned to be implemented in various GIS environments, which offer a wide range of spatial functions [11, 12, 13, 14] and new database tools [15, 16, 17]. Thus, the dynamic modeling can be shared together with spatial modeling that offers exploratory spatial data analysis, advanced methods of spatial interpolation based on geostatistical methods, digital terrain modeling extended by complex surface analysis, network optimization methods, raster algebra for creation of multi-criteria analysis, conversion tools for a wide range GIS/non-GIS data formats, and integration of spatio-temporal data to the spatial databases for sharing digital map layers by a number of other software tools dedicated to the case oriented tasks.

Another way of the AVSim implementation represents numerical tools included in the simulation languages (Advanced Continuous Simulation Language-ACSL) or high-level languages with interactive environment that enables to perform computationally intensive tasks faster than with traditional programming languages (MATLAB/Simulink, Mathematica, Stella) [18, 19, 20, 21, 22, 23, 24, 25]. In spite of that the numerical tools can solve time consuming numerical calculation by a number of numerical simulation methods, there is need to develop the conversion tools for sharing data with GIS.

In addition to the standard desktop application, such as ArcGIS for spatio-temporal modeling or MATLAB for dynamic modeling, new computer approaches offer to use modeling tools based on the client-server software architecture. Basically, client-server means the relationship between two computer programs in which one program, the client program, makes a service request to another, the server program. For example, standard networked functions, such as web access and database access, are based on the client-server model. A web browser is a client program at the user computer that may access information at any web server in the world. In case of spatio-temporal modeling, a GIS web browser client program in the user computer can forward the user request to a GIS web server program at the computer center. That program may in turn forward the request to its spatio-temporal database server extended by numerical modeling tools for the model simulation. After the simulation tasks, the result is returned to the web browser client in the user personal computer, which displays the information. Nowadays, there are a few software systems based on client-server architecture that can support spatio-temporal modeling task. For example, ArcGIS server is able to connect people with the geographic information they need. Users may use ArcGIS Server to distribute maps and GIS capabilities via Web mapping applications and services to improve internal workflows. But, it can also take layers with modeling outputs and publish them as Web services, giving Web application users powerful tools for data and surface exploration, such as display biohazard release and predicting minimal impacts on living environment by using optimal remediation procedures.

As an example, a schema of the computer network that can support all the described alternatives for spatio-temporal environmental modeling is in Figure 13. While the AVSim implementations are described in the GIS desktop environment (ArcGIS/ArcView) and in the standard simulation tools (MATLAB), the computer network based on the client-server architecture shows other more complex solutions. A solution based on ArcGIS server can serve the map service, geodata service, and image service. The GIS server hosts user resources, such as maps, globes, images services, and exposes them as services to client applications. The server object manager (SOM) connects to one or more server object containers (SOC), and manages them. There can be multiple SOM and SOC running on different machines. The data server contains the GIS resources (map documents, globe documents and raster datasets) for publishing as services on the GIS server. The ArcGIS Image server extension acts as the request broker for ArcGIS Image Server connections. The Web server hosts Web applications and Web services that use the resources running on the GIS server. On the other side, Web client applications can be Web, mobile, and desktop applications that connect over HTTP to ArcGIS Server Internet services.

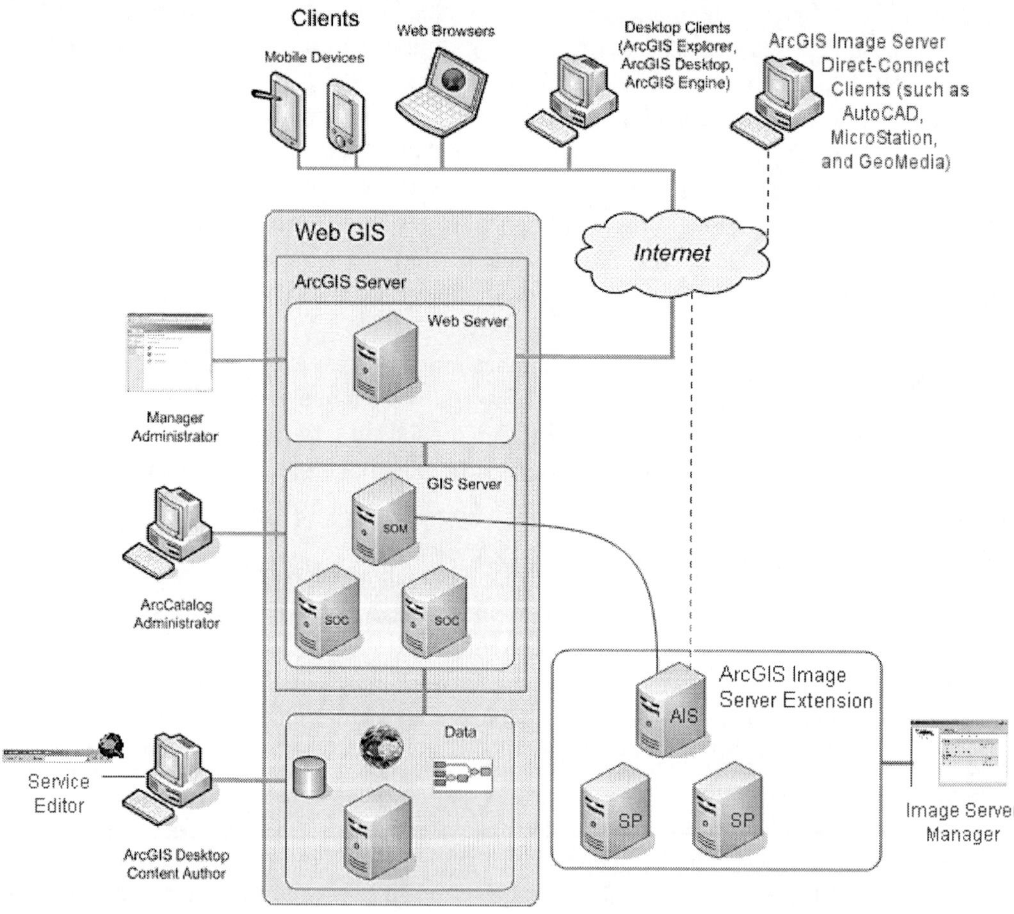

Figure 13. The computer network that can support all the described alternatives for spatio-temporal environmental modeling based on the client-server architecture supported by ESRI's GIS.

The described modeling of the surface water pollution can be used for a wide range of similar environmental problems. The demonstrated spatio-temporal modeling in the GIS environment offers to share other data and methods that are valuable for risk assessment. The new approaches based on client-server technology implemented in the GIS environment offer more efficient ways of spatio-temporal modeling for risk assessment and decision-making processes. They can integrate powerful computation tools installed in the research centers together with the terrain mobile computational systems that are able to capture actual information and provide online measurements. Thus, after the pollutant release, the remediation procedures can be optimized in order to minimize the impacts on living environment and human health. In case of the GIS environment, sharing of data from GPS and remote sensing represents other possibilities how to solve environmental problems on a higher professional level.

ACKNOWLEDGMENT

The presented research was carried out in the framework of the GIS Laboratory research project supported by the Ministry of Education, Youth and Sports. The partial results were presented at the 84th Annual Meeting of Ecological Society of America in Spokane in 1999. The cooperation of the authorities from the Vltava Basin is gratefully appreciated.

REFERENCES

[1] Jørgensen, S.E. and Bendoricchio (2001). *Fundamentals of Ecological Modelling.* Amsterdam: Elsevier, 3rd ed.

[2] Davis, L.R., (1999). *Fundamentals of Environmental Discharge Modeling.* London: CRC Press.

[3] Logan, J.D. (2001). *Transport Modeling in Hydrogeochemical Systems.* New York: Springer.

[4] Hannon, B. and Ruth M. (1994). *Dynamic Modeling.* New York: Springer.

[5] Matejicek, L. (2005). Spatial modelling of air pollution in urban areas with GIS: a case study on integrated database development. *Advances in Geosciences 4*, 63-68.

[6] Matejicek, L. (1999). Spatio-temporal modelling in the GIS environment: Dynamic processes of water pollution in streams. In: G.R. Larocque (Ed.), Ecomod, October 1999, pp. 16-17.

[7] Matejicek, L. (1999). Modelling of groundwater flow networks in the desktop GIS environment. In: L. Oxley, F. Scrimgeour (Eds.), Modelling the Dynamics of Natural. Agricultural, Tourism and Socio-economic Systems. University of Waikato: Modelling and Simulation Society of Australia and New Zealand, pp. 55-60.

[8] Matejicek, L. and Tlapakova L. (2002). Modeling in the GIS environment: dynamic processes of water pollution. *Acta Universitatis Carolinae Environmentalica 16*, 47-62.

[9] Spanou, M., Chen, D. (2000). An object-oriented tool for the control of point-source pollution in river systems. *Environmental Modelling & Software 15*, 35-54.

[10] Tlapakova L. (2002). Modeling of the pollution propagation in the Zelivka river basin. *Acta Universitatis Carolinae Environmentalica 16*, 37-45.

[11] Goodchild, M.F., Steyaert, L.T. and Parks, B.O. (1996). *GIS and Environmental Modeling: Progress and Research Issues.* Fort Collins: GIS World, Inc.

[12] Zeiler, M. (1999). *Modeling Our World.* Redlands, California: ESRI Press.

[13] Maguire, D.J., Batty, M. and Goodchild, M.F. (2005). *GIS, Spatial Analysis and Modeling.* Redlands, California: ESRI Press.

[14] Mitchell, A. (2005). *The ESRI Guide to GIS Analysis. Volume 2: Spatial Measurements & Statistics*. Redlands, California: ESRI Press.

[15] Rigaux, P., Scholl, M. and Voisard, A. (2002). *Spatial Databases with Application to GIS*. San Francisco, California: Morgan Kaufmann Publishers.

[16] Arctur, D., Zeiler, M. (2004). *Designing Geodatabases: Case Studies in GIS Data Modeling*. Redlands, California: ESRI Press.

[17] MacDonald, A. (2001). *Building a Geodatabase*. Redlands, California: ESRI Press.

[18] Holzbecher, E. (2007). *Environmental Modeling Using MATLAB*. Berlin: Springer.

[19] Littlefield, B.L. and Hanselman, D.C. (2004). *Mastering Matlab 7*. USA: Prentice-Hall.

[20] Dabney, J.B. and Harman, T.L. (2003). *Mastering Simulink*. USA: Prentice-Hall.

[21] Trauth, M.H. (2007). MATLAB Recipes for Earth Sciences. Berlin: Springer, 2nd ed.

[22] Marchand, P. and Holland, O.T. (2007). Graphics and GUIs with MATLAB. Florida: CRC Press, 3rd ed.

[23] Costanza, R. and Voinov, A. (2004). Landscape Simulation Modeling: A Spatially Explicit, Dynamic Approach. New York: Springer.

[24] Ford, A. (1999). Modeling the Environment: An Introduction To System Dynamics Modeling Of Environmental Systems. Washington DC: Island Press.

[25] Deaton, M.L. and Winebrake, J.J. (1999). Dynamic Modeling of Environmental Systems. New York: Springer.

In: Environmental Modeling with GIS
Editor: Lubos Matejicek, pp. 33-54

ISBN: 978-1-60876-363-4

Chapter 3

MODELING OF WATER POLLUTION IN RIVER BASINS WITH GIS

Lubos Matejicek

Institute for Environmental Studies, Charles University in Prague, Faculty of Natural Science, Benatska 2, 128 01, Prague, Czech Republic

ABSTRACT

The increasing importance and awareness of water quality in river basins resulted in the establishment of wide-ranging monitoring networks. Long-term measurements of physical-chemical parameters, hydrological and meteorological observations, mapping of basins with GIS, GPS and remote sensing requires development of advanced tools for data analysis and modeling in the frame of an information system. A few strategies to integrate spatially distributed environmental data and models are presented with use of the recent development in satellite technology, which can support classification of basin surface to identify potential sources of pollution. Data analysis and compartment models deal with basin characteristics extended by land cover attributes, which are in addition to standard terrain measurements estimated by processing of satellite images from Landsat 7 TM and ASTER. A number of indexes, for example NDVI is used to set model parameters. A case study focused on water pollution from nitrates demonstrates accumulation and transport of pollutants from point and non-point sources. The dynamic models are integrated into GIS from data linkage through shared data files to building models as extensions into fully functional GIS. Other way is represented by development of a standalone software application that covers GIS and modeling functionality. As an example of modeling of water pollution in the GIS environment, a computer program developed with the MapObjects programming library documents abilities to manage spatial data, calculate dynamic models and display results together with the storage of data in spatial formats or database.

Keywords: Spatio-temporal modeling; surface water pollution; GIS.

1. Introduction

Ecological modeling and Geographic Information Systems (GIS) combine the functionality of two independently designed systems. The power of ecological modeling sums up abilities to describe dynamic phenomena by mathematical equations. GIS can manage a wide range of spatial data formats including preprocessing and post-processing of model inputs and outputs. In such combinations, GIS performs functions that are mostly focused on data management in the frame of a spatial database, coordinate transformation and projection change, conversion between spatial data formats and structures, raster algebra, vector operations and visualization of model results. Preprocessing represents data preparation to assemble all initial conditions and parameters of dynamic models. Post-processing usually serves as data conversion of model outputs to display and spatially analyze results. A series of data structures has been proposed to build connection between geospatial and temporal data, which is for example illustrated in the chapter of the book GIS and Environmental Modeling by Fedra [1]. Affiliated examples document various ways of data exchange between GIS and modeling of water, soil and air pollution. Despite obvious diversity, it is possible to identify common themes that are focused on spatial database development, data linking between ecological modeling and GIS, development of preprocessing and post-processing tools in the frame of GIS. Integrating hydrologic modeling and GIS, which involves connecting spatial data describing the physical environment with hydrologic process models, introduces Maidment [2]. His Arc Hydro data model for water resources opens the way to building hydrologic information systems that include spatial and temporal water resources data to support data analysis and dynamic modeling. The Arc Hydro provides an environment for integrating of geospatial and time series data for water resources with simulation models for water flow and quality. A variety of application modeling programs can be accessed through existing built up interfaces. In spite of GIS interoperability, a series of simulations has to be carried out by standalone software applications. It is caused by a huge range of phenomena solved in the frame of water pollution modeling. It also documents contributions in the book edited by Wrobel and Brebbia [3] and many presentations at conferences shared by modeling and GIS societies [4, 5].

2. Methods

Without the understanding of the physical, chemical and biological phenomena, it is not possible to formulate models and without models there cannot be quantification of the effects of polluting discharges. The past decades have seen a marked improvement in our understanding of the physical, chemical and biological phenomena in basins and an increase in our ability to model the principal processes of wastewater discharges. But the limitation has been indicated in precise description of spatial interactions, standard data formats of simulation tools and integrating new methods of terrain measurements, remote sensing data and global positioning systems.

2.1 Integration Strategies of Dynamic Modeling and GIS

Modeling of transport and accumulation of pollutants in basins can be carried out by a number of simulation tools. Simulation solved by finite-element and finite-difference numerical models, which are represented usually by two-dimensional dynamic equations, corresponds to the description of transport and accumulation of pollutants. Due to amount of numerical calculation, there is a number of standalone software programs in water pollution modeling. Some of them are for example referenced by a book edited by J0rgensen [6]. In this case, the numerical models do not presuppose storage of data in any GIS formats that can support spatial analysis. Before and after simulation, data are transported into GIS in the frame of preprocessing and post-processing. In the basic case, two separate systems (the GIS and the dynamic model) constitute modeling tools, which is illustrated in Figure 1. Mostly, the dynamic model obtains some of its parameters, initial conditions and data inputs from the GIS and produces some of its outputs in a format that can be transported into the GIS. Different user interfaces need to be managed, which can be error prone if a significant amount of manual tasks is involved.

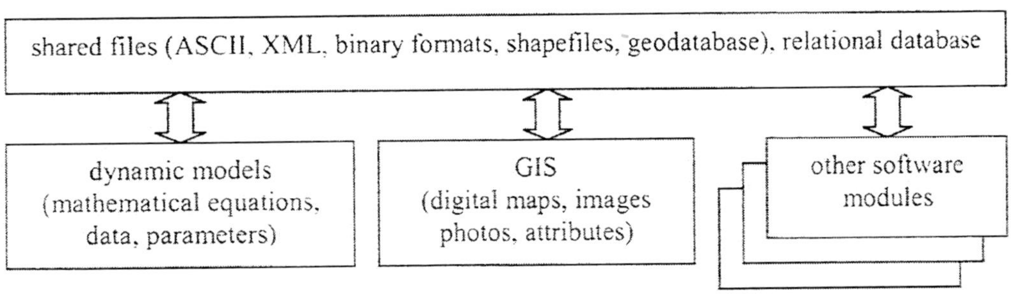

Figure 1. Dynamic modeling extended by GIS analysis through data exchange.

Simplification of two-dimensional dynamic equations into the form of ordinary differential equations, which describe accumulation of pollutants in relatively homogeneous parts of basins, represents more general models that can be solved by many standard mathematical software programs. Numerical algorithms for solution of model equations can be also implemented in the frame of the GIS environment. One way to achieve this task is to use a higher-level application language built into a GIS development kit. ArcObjects programming library in the frame of ArcGIS can represent one of such approaches [7]. The programming library contains all the functions used to develop GIS software applications like ArcCatalog, ArcMap, ArcTools, etc. Other software applications can be developed as extensions of existing modules or as standalone modules. Modeling of water pollution by integrated simulation tools in the frame of GIS is illustrated in Figure 2. In this case, GIS offers robust tools to construct spatial data models extended by topological rules. Moreover to standard vector and raster formats, integration of Triangulated Irregular Networks (TIN) for building of Digital Elevation Models (DEM) and data from remote sensing is created to provide more complex analysis [8, 9].

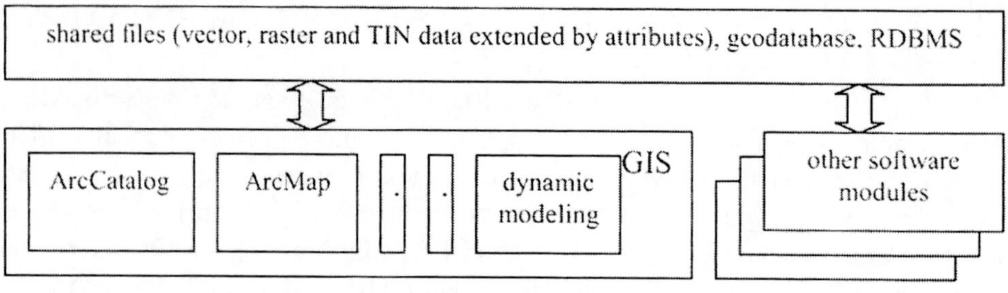

Figure 2. Dynamic modeling in the framework of GIS-ArcGIS.

Wide functionality for development of spatial data models and analysis database management and integration of extensions to solve dynamic models of water pollution in the GIS environment requires an additional user experience. While efficient for fast prototyping, using predefined components often needs to modify or reformulate original aims. So, another alternative is to use standalone developer kits that provide both GIS functionality as well as components for simulation of dynamic models. Figure 3 shows a project dedicated to programming of a software application for modeling of water pollution extended by GIS analysis.

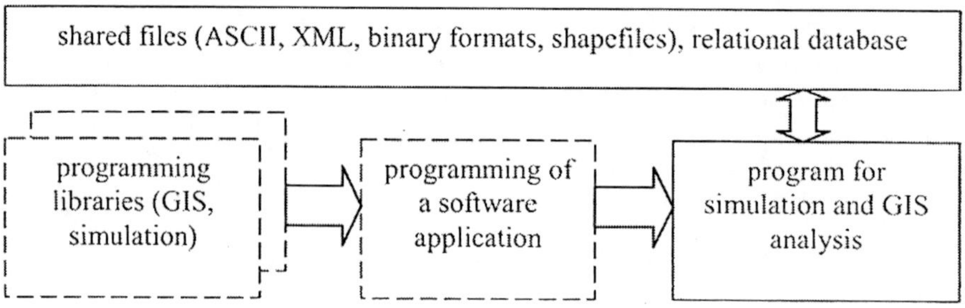

Figure 3.Development of a standalone software application for modeling of water pollution.

2.2 Integration of Spatial Data from Aerial and Satellite Images

The spatial and temporal categorizations of land cover represent one of fundamental data sets for modeling of water pollution in basins. There is a number of methods for deriving of land cover distribution from images in remote sensing. Different remote sensors are capable of measuring various regions of the light spectrum. Aerial photography and satellite images are being used more and more to detect and inventory the severity of water pollution sources in basins. Point sources can be localized as the discharge of industrial wastes into bodies of water. Non-point sources cover relatively larger areas, which are represented by fertilizer runoff from agricultural fields, sediment runoff following heavy rainfall or the leaching of sewage over large areas. Some sources of pollution can be detected directly on the images, some indirectly and a remaining part by terrain measurements. The direct detection indicates a visible change in tone, color or texture of the water as imaged. The indirect detection is the result of detectable changes in the environment caused by water pollution, which can be represented as the killing of vegetation or aquatic life. Mostly, the influential part of sources has to be detected by terrain measurements or another method of identification. In case of massive water pollution, materials suspended in bodies of water cause the image tones to be lighter because more energy is reflected and less energy absorbed by the water. Another type of pollution like for example heat pollution is detectable by thermal scanners.

In the frame of complex studies of matter and energy flows in water ecosystems, the need of surface categorization is necessary to estimate accumulation and transport properties of surrounding areas. Moreover surface objects detected from the visible regions of the electromagnetic spectrum, the near-infrared, shortwave-infrared and thermal-infrared regions can be used for forestry and agricultural land cover analysis. From a wide range of the aerial and satellite sensors, the images from Landsat Thematic Mapper (TM) and Advanced Spaceborne Thermal Emission and Reflection Radiometer (ASTER) are used to study land cover in the basins. TM data (3 visible and 4 infrared/thermal spectral bands, 30 m spatial resolution and 185 km imaging swath) can be classified into a few main classes that include surface water, forested areas. agricultural areas and bare soils. Classification methods, which are based on a segmentation of the images into regions with similar radiometric properties, are often used as a first step for mapping. In addition to classification, a number of indexes can be calculated to quantify land cover characteristics. Commonly used vegetation descriptors are the Vegetation Indexes. The Normalized Difference Vegetation Index (NDVI) is calculated on a pixel-by-pixel basis by subtracting the intensity of the infrared band from the visible red band:

$$NDVI = \frac{R_{nir} - R_{red}}{R_{nir} + R_{red}}$$

where the parameter R_{red} marks the red reflectance and the parameter R_{nir} the near-infrared reflectance. Resulting index values of a pixel typically range from around 0.1 to 0.6 for vegetation. The higher value means the denser vegetation. Rocks, bare soil and man-made objects produce the index value near 0.0. Other modifications of the NDVI together with an approach to land cover classification with TM data are discussed for example by Stefanov [10]. In comparison to the size of a TM scene, classifications using ASTER data (2 visible and 13 infrared/thermal spectral bands, 15-90 m spatial resolution. 60 km imaging swath) are focused on smaller areas. A number of bands, which are scanned by three separate optical subsystems (4 bands in the visible and near-infrared radiometer- VNIR, 6 bands in the shortwave-infrared radiometer-SWIR and 5 bands in the thermal radiometer-TIR), offers more elaborative evaluation [11, 12, 13]. The thermally emitted radiance, which depends on the surface temperature and the surface emissivity, characterizes the composition of the surface, especially moisture content.

TM and ASTER data together with GIS layers and terrain measurements serve for classification of a basin surface and calculation of indexes, which are used for estimation of potential accumulation and transport of water pollution in a basin. After spatial analysis all the data in GIS, diffusion and advection parameters in dynamic models can be adjusted in the appropriate scales [14, 15, 16].

2.3 Integration of Spatial Data from GPS

Global Positioning Systems (GPS) evolved from dedicated military systems have been utilized in numerous civil applications. The fully operational GPS includes 24 or more active satellites approximately uniformly dispersed around six circular orbits with four or more satellites each. Theoretically four or more GPS satellites can be used to determine an observer's position including an elevation anywhere on the earth's surface 24 hours per day. The interactions together with other earth's stations, corrections and post-processing of data are described by Grewal [17]. Recent advances in mobile computing represented by mobile devices and ESRI's ArcPad make possible to automate a number of procedures in addition to collection of positions in the basin. Basic statistical methods and smaller dynamic models can be included into mobile computing in the frame of GIS and used to estimate new features and impacts of man-made activities. After measurements and estimates of parameters, all data can be transferred into a central spatial database that is shared by other GIS tools [5]. A basic schema of a mobile GIS for the data collection in a basin is illustrated in Figure 4.

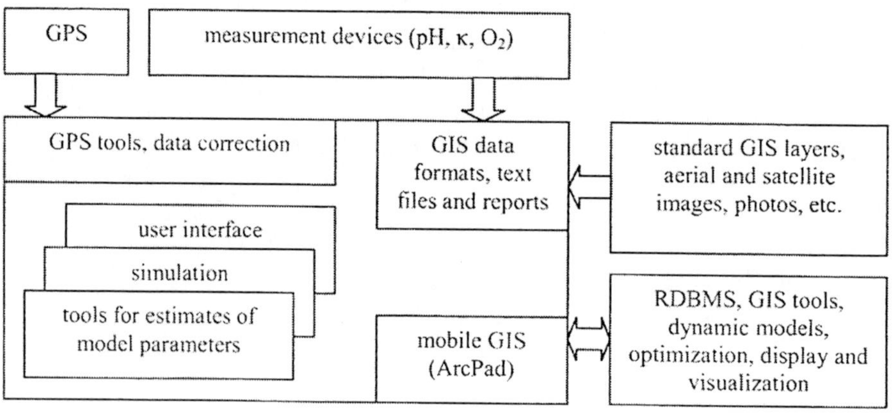

Figure 4. Data collection supported by GPS.

2.4 Integration of 3D Spatial Data

Modeling of water pollution deals with physical-chemical data of water streams and surrounded areas. A size of objects on the surface together with elevations besides other data is needed to model hydrodynamic phenomena. In addition to terrain measurements, the reliable quantitative information about spatial data of the surface can be also obtained from aerial and satellite images. A series of methods, which include building of 3D data models from aerial and satellite images, is one of the main topics of photogrammetry. Because of advanced technology and increased availability, the images become an important component for all natural resources disciplines, as well as for land-use planning together with water pollution studies. Principles of photo interpretation from many different kinds of remote sensors like for example aerial and satellite images, radar and lidar and digital cameras, which acquire data without making physical contact with the sources, are described by Paine [18].

In the frame of modeling of water pollution, a precise DEM together with surface objects like water barriers, buildings and vegetation is needed for drainage analysis in the surrounded areas of water streams and estimating of hydrodynamic parameters. Drainage is the flow process by which water moves from the points that fall onto the landscape down to streams, then to rivers. Drainage flows downhill so the pollution of the land surface from point and non-point sources is a function of its shape. GIS tools contain functions to accomplish the drainage area and the stream network delineation from DEM. In spite of it, manual editing of drainage boundaries may be necessary, especially in case of regions with very flat terrain.

Association of drainage areas with individual streams subdivides the basin landscape into a large number of elementary drainage areas. In this case, the path of a raindrop and pollution from point and non-point sources on the land surface from anywhere in the drainage area will be connected with its associated stream. So the DEM is used to subdivide the basin on subbasins, which are represented by compartments in the model of water pollution [19].

2.5 Integration of All the Spatial Components in the Framework of GIS

Spatial data from aerial and satellite images, GPS data extended by physical-chemical measurements and 3D terrain models can be integrated together with dynamic modeling in GIS. A wide range of data formats and the support of spatial analysis and modeling enable to develop projects focused on water pollution in basins. Special programming languages are used to customize GIS for particular applications, such as to create GIS preprocessing and post-processing modules for simulation of water pollution models or building standalone applications. Figure 5 illustrates development of a GIS project that can support modeling of water pollution. In the frame of the project, data from remote sensing are processed to obtain GIS layers of point and non-point sources of pollution, classification of the basin surface area and DEM. GPS data extended by terrain measurements from mobile devices complement existing records that include monitoring of water pollution in the basin and meteorological data. Actual spatial data are represented by other map layers, which extend existing digital maps and images managed by local and state authorities. In addition to integration of dynamic modeling on various levels, other external connections to a relational database and Internet services enable sharing of other data.

3. A Case Study of Basin in the Neighborhood of Praha Focused on Nitrates

The increasing importance and awareness of water pollution in the basins resulted in development of monitoring networks on the level of state authorities and agencies. The significant part of the water pollutants like heavy metals, organic compounds, total phosphorus, ammonia, nitrates and pesticides is monitored by automatic stations and terrain

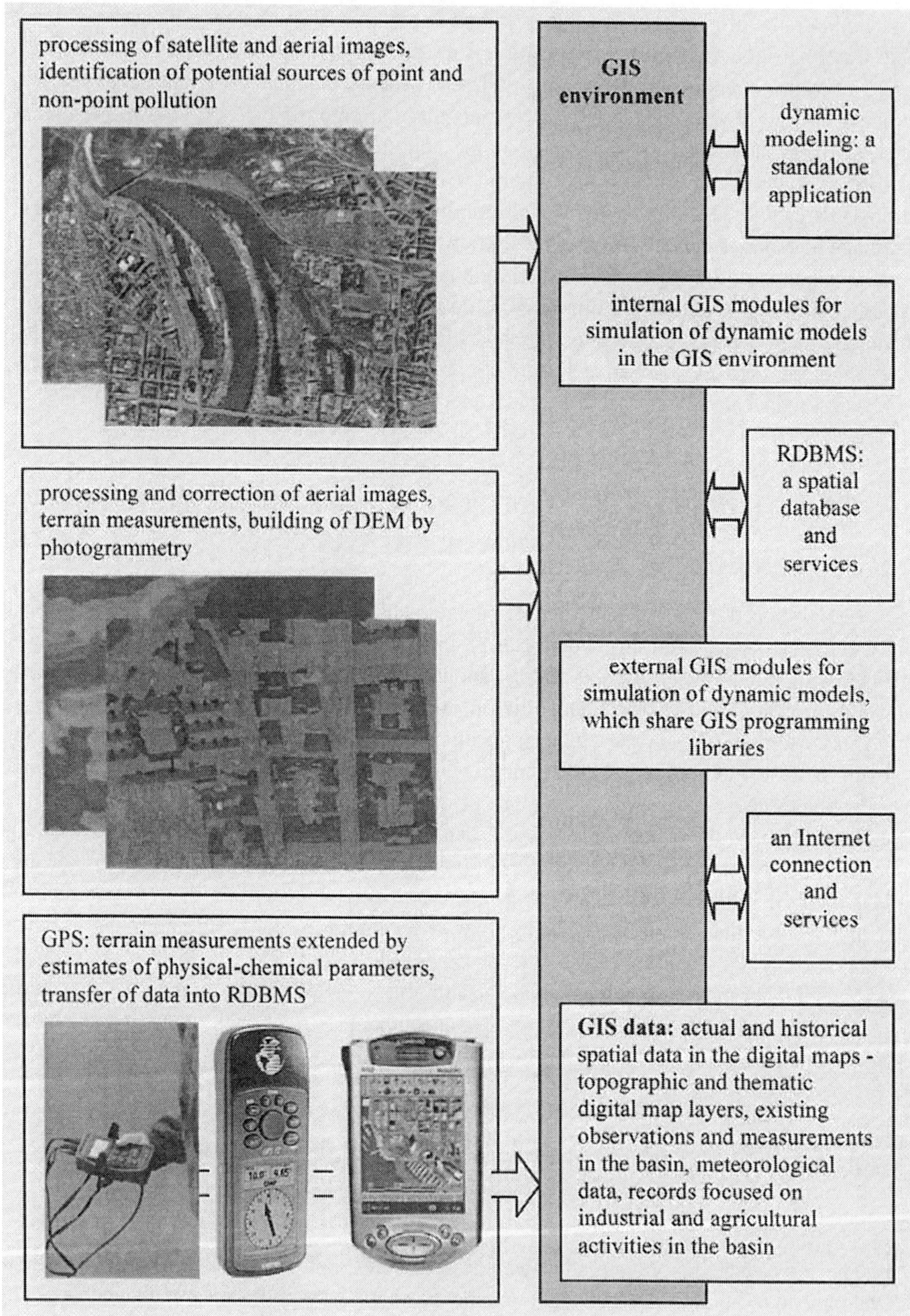

Figure 5. A GIS project for modeling of water pollution in a basin.

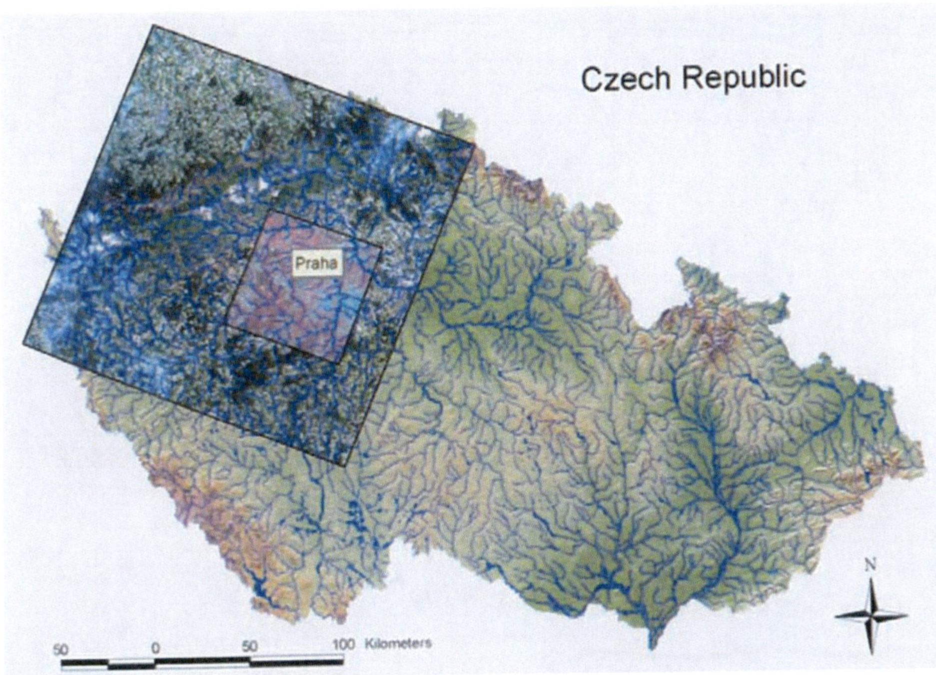

Figure 6. Display of satellite images from Landsat 7 TM (bands: 1, 2, 3) and ASTER (bands of VNIR: 1, 2, 3N) across the studied area on the territory of the CzechRepublic in the neighborhood of Praha from ArcCR 500 [©ArcCR 500: digital geographic database 1: 500 000, ARCDATA s.r.o.].

Figure 7. The detailed view of the basins in the neighborhood of the capital Praha (Prague) composed by the image from Landsat 7 TM (bands: 1, 2, 3).

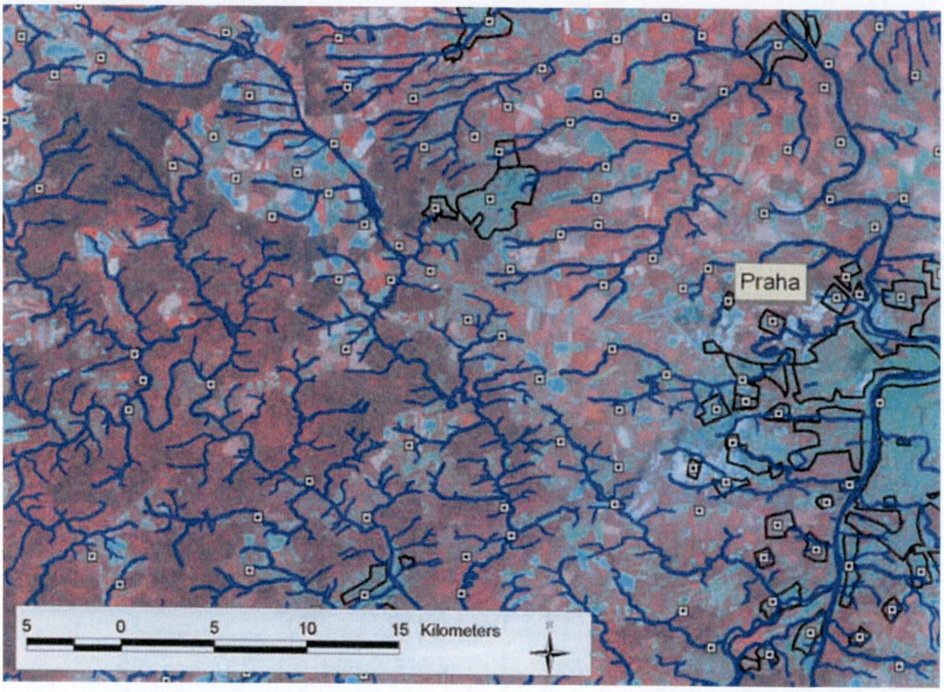

Figure 8. The detailed view of the basins in the neighborhood of the capital Praha (Prague) composed by the image from ASTER (bands: 1, 2, 3N).

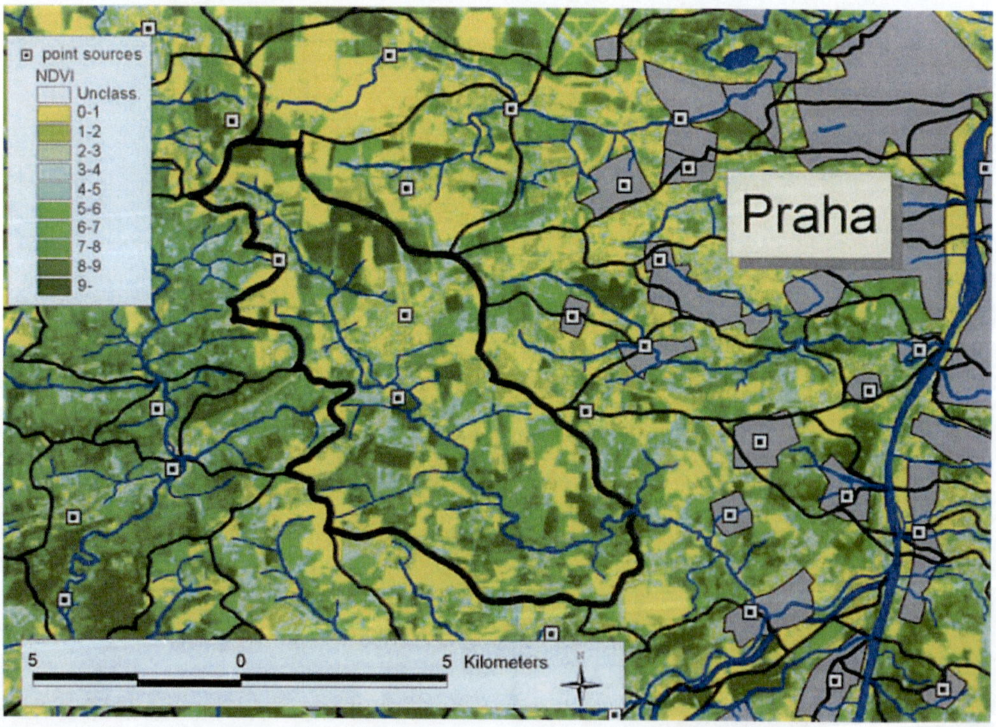

Figure 9. The normalized difference vegetation index (NDVI) calculated on a pixel-by pixel basis from the red and near infrared parts of the light spectra (the marked area represent a hydrological unit, which is used as a section in the compartment model).

measurements. Figure 6 shows rivers in the Czech Republic. The river layer is complemented by DEM and satellite images, which delimit a part included into a GIS project. The larger satellite image demonstrates data from Landsat 7 TM captured in May 2000. The image composition is carried out using blue (1), green (2) and red (3) bands of the visible part of the light spectra. Other infrared bands (4, 5, 6, 7) are used for studying and classification of vegetation and soil surface in the surrounded part of the basins. The ASTER image captured in May 2002 is situated inside the Landsat 7 TM image. In this case, two bands (1, 2) from a visible part of spectra and one near infrared band (3N) are used for image composition. Infrared bands assist in the frame of mapping of vegetation, soil, minerals and rocks on the surface. All bands in satellite scenes complemented by classification and mapping layers in the basins represent data used for identification of potential sources of water pollution and setting of parameters in the dynamic models.

The more detailed view of the basins covered by the Landsat 7 TM scene is illustrated in Figure 7. The ASTER scene is displayed in Figure 8. Both the satellite images with 30 m spatial resolution enable to recognize parts covered by forest (ASTER: dark red), agricultural soils (ASTER: mostly light shades of red), bare soils (ASTER: mostly cyan) and urban areas (ASTER: cyan). The images are completed by a line layer of the rivers and a point layer of urban settlements. The satellite images form just a part of the GIS project. Other important parts are represented by standard GIS layers provided by mapping agencies and hydrological data monitored by state authorities. In case of water pollution modeling, fundamental subsystems constitute data analysis and dynamic modeling complemented by map locations of point and non-point sources of pollution. The map layer of settlements and cities is used for the estimate of the point sources of pollution. The balance of non-point sources of pollution is built on the NDVI and terrain measurements. The NDVI image calculated for the studied area is showed in Figure 9. The NDVI is divided into 11 classes. The dark green areas with a higher value of the NDVI are mostly situated on the left side of the scene, which represents the forested upper part of the inflow streams to the main basins. The yellow and light green areas with a low value of the NDVI cover urban, industrial and agricultural areas where erosion processes in dependency on slope and soil characteristics can form potential non-point sources of pollution. These data together with terrain measurements are used for setting of parameters, which summarize the total influence of the surrounded area on water pollution in each section of the basin. The particular setting of model parameters focused on nitrate pollution is described by Matejicek [20].

3.1 Model Description of Water Pollution

The nitrate transport depends on a huge range of factors, which include particular agricultural and industrial management in each basin, land cover, meteorological and hydrological conditions, processes in soils, sediments and surface water as described by Norton and Fisher [21]. The reduction of nitrate loads can be correct by using knowledge about nitrate emissions, flow paths and turnover processes together, which are converted into the description of dynamic models. Dynamic models simulate all these processes in spatial and temporal dimensions, which can save additional measurements. After validation and verification of dynamic models, prediction of nitrate concentrations in the basins can be carried out to support decisions in environmental management and planning. A model scale on the level of satellite images and terrain measurements enables to study the major nitrate paths and transformations in each section of the basin. The potential nitrate sources, tanks and transformers are estimated by using of classification and categorization of surface objects (rivers, lakes, forest and arable land) from GIS. Mostly, the arable land and large urban areas are considered as the main non-point sources of nitrate pollution. Point sources of nitrates are located in industrial zones and smaller residential areas.

Over a few decades, a series of models has been built to solve pollution of water ecosystems. Many of them are administrated by EPA or sold by private companies. Examples of compartment dynamic models calculated by STELLA are described by Hannon and Ruth [22]. Environmental modeling linked to GIS is introduced in the book GIS and Environmental Modeling: Progress and Research Issues by Moore [23]. In spite of a number of models focused on dynamic phenomena, the common research, which combines the power of environmental modeling software with GIS spatial analysis, has been quite successful over the last decade.

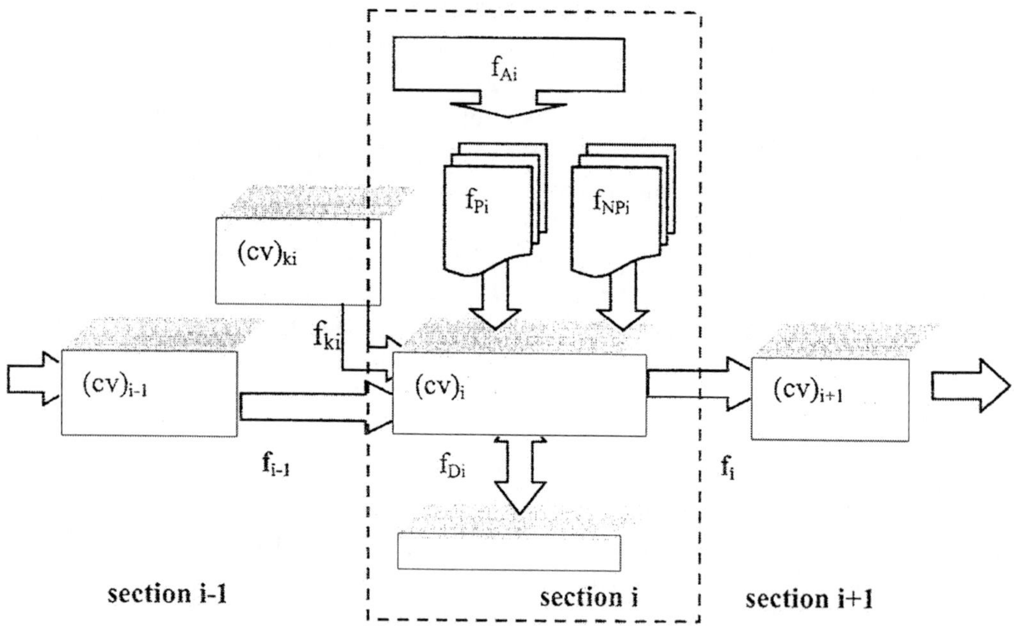

Figure 10. The schema of nitrate accumulation and transport in a section of the basin.

In case of compartment models, which can reduce the amount of calculation in comparison to standard diffusive-advection models, each of the river basins is divided into the compartments-sections based on independent hydrological units. The physically based time continuous accumulation and transport of nitrates can be built on the schema in Figure 10. Accumulation of nitrates is illustrated by boxes that contain the amounts of nitrates $(cv)_i$, where c marks the concentration of nitrates in the volume v of the i-th section. The input flows of nitrates into the i-th section represent all the local inflows f_{Ki}, the point and non-point sources f_{Pi}, f_{NPi}, the atmospheric deposition f_{Ai} and the interactions with sediment and biota f_{Di}.

The transport of nitrates into the *i*-the section from the upper stream section *i-l* is caused by the input flow f_{i-1}, which can be estimated as $(cv)_{i-1}q_{i-1}/v_{i-1}$. The decrease of the amount of nitrates from the *i*-th section is represented by the output flow of nitrates f_i into the downstream section *i*+1, which can be estimated as $(cv)_i q_i/v_i$. The variable q_i is the appropriate water flow. In spite of similar interactions in reservoirs, the amounts and flows of nitrates are diametrically different. Individual subsections have to be created to include differences In denitrification, sedimentation and biological uptake or increase due to mineralization.

A complicated part of the model development is comprised by estimates of nitrate inputs into the streams from non-point sources of pollution. A number of models attempt to solve accumulation and transport of nitrates from surrounded areas [24, 25]. In case of this model, the estimate of nitrate pollution from non-point sources is built on hydrological and meteorological characteristics, soil properties and land cover of each section, which is particularly obtained from satellite images.

3.2 Specification of Input Data

Water quality in the studied basins has been monitored for a few decades by state authorities and agencies (Figure 11, 12). More detailed measurements are available for shorter periods in the frame of various research projects. As an example, time series of physical-chemical variables are demonstrated in the period 1985-2002. The time interval of individual measurements is a month. Each time series is complemented by a series of moving averages to display year average values. Samples represent a part of all data, which are measured and administrated by the state authorities of the basin Vltava. In the majority of monitoring points, the nutrient concentrations are highly variable, which can alert on water pollution and erosion in the basins. Data are comparable with those measured and estimated for other subsidiary basins dominated by industry and agriculture. The basic task of modeling of water pollution is the dynamic model, which is able to simulate measured time series and after that forecast trends.

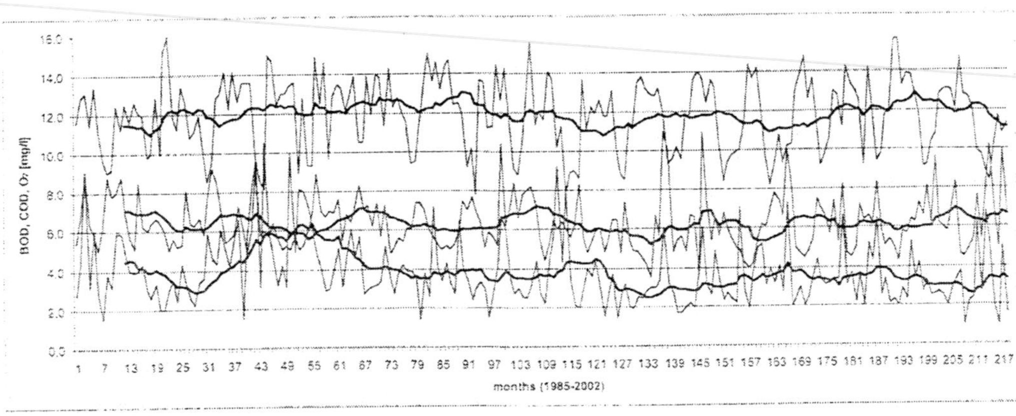

Figure 11. An example of measured time series of O_2 (upper line), BOD (middle line) and COD (bottom line) on the monthly time step.

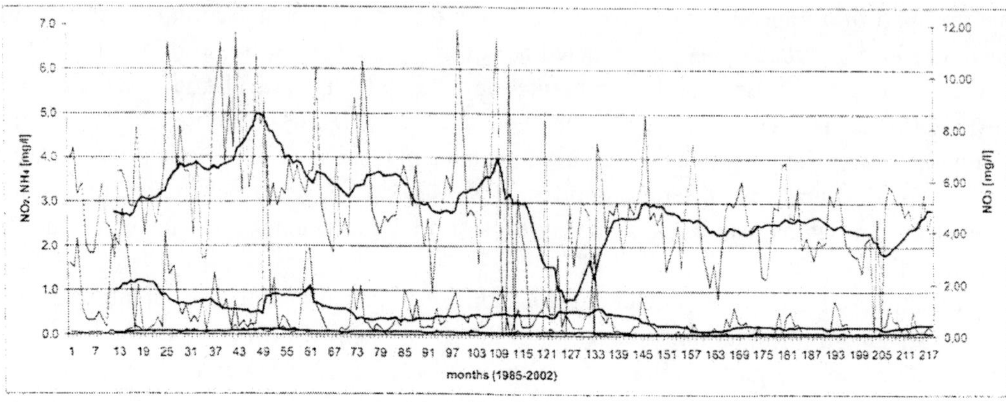

Figure 12. An example of measured time series of NO_2 (upper line), NH_4 (middle line) and NO_3 (bottom line)scaled by a secondary axis) on the monthly time step.

3.3 Description of the Dynamic Model

Accumulation and transport of nitrates in the i-th section of the basin can be described by the ordinary differential equation:

$$\frac{d(cv)_i}{dt} = f_{i-1} + \sum_{k=1}^{m} f_{ki} - (cv)_i \frac{q_i}{v_i} + f_{Pi} + f_{NPi} + f_{Ai} - f_{Di}$$

where the variable f represents the individual nitrate flow with correspondence to the schema in figure 10. The variable $(cv)_i$ marked as the nitrate concentration e and multiply by the volume v is the amount of nitrates in the i-th section. The term $(cv)q_i/v_i$ represents the nitrate outflow f_i from the i-th section. The nitrate inflow f_{i-1} is estimated in a similar way as an outflow from the upper section. The compartment model assumes homogeneous mixing of nitrates in the volume v_i, which is estimated as the amount of water in the river bed or in the reservoir. The homogeneous mixing sums up an approximation of real processes. The smaller or less indented the water volume the better the approximation of hydrodynamic processes. Other nitrates inflows from surrounding basins are represented by the sum of f_{ki}. In case of the point sources of water pollution, the nitrate inflows f_{Pi} can be estimated by terrain measurements or from annual reports, which are carried out by state authorities for some point sources in the frame of water quality monitoring. Contributions of nitrates from nonpoint source depend on a number of factors, which cover meteorological, hydrological and geochemical processes. The terrain measurements and estimations of nitrate outflows from various types of surrounding areas (forest, meadows, arable land and industrial zones) carried out in the frame of local projects are used for an approximation in the dynamic model [26]. The important moment in this research represents the use of the satellite data for classification and categorization of various types of the surface. A total nitrate inflow from non-point sources of pollution is estimated on the basis of a proportion of land cover and terrain morphology. The initial setting of inflows from non-point sources is proposed to be specified more precisely during the simulation and verification phases of modeling. This is also the case of denitrification flows f_{Di} that are seasonally variable. The nitrate flow f_{Di} covers a number of abiotic and biotic processes in the water ecosystem [20].

3.4 Integration of the Dynamic Model with GIS

A few strategies of integration of dynamic modeling and GIS are described in the previous parts. The proposed compartment model is implemented as a standalone software application, an extension of an existing GIS that include a development programming kit and a software application based on programming libraries that support functionality of dynamic modeling and GIS. Observed time series at monitoring points, map layers with attributes, aerial and satellite images, meteorological and hydrological characteristics need to build a spatial database in the frame of GIS. In case of the spatial database, a variety of interfaces exists for accessing all the data, including ArcGIS and standard office applications like for example Microsoft Access or Microsoft Excel. Likewise the interfaces HEC-GeoHMS and HEC-GeoRAS for hydrologic/hydraulic simulation models HEC-HMS and HEC-RAS (the Hydrologic Engineering Center of the U.S. Army Corps of Engineers) or EPA Basins program for water-quality simulation were developed in the ArcView 3 GIS environment, the partial task focused on modeling of nitrate flows in the basin can be carried out in the frame of the ArcGIS or the MapObjects programming library. A few ways of integration of dynamic modeling tools with GIS are introduced for the described model.

3.5 Spreadsheet Tools and GIS

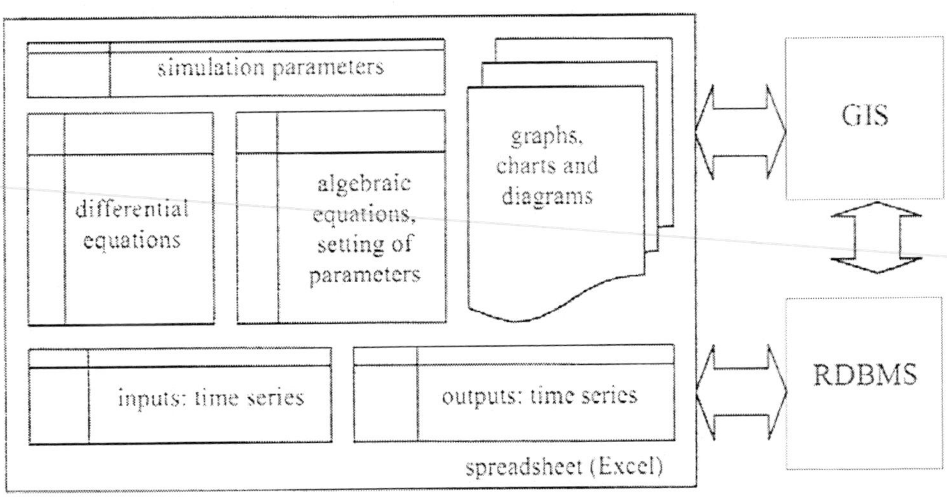

Figure 13. Calculation of the dynamic model in the environment of a spreadsheet and data connection among the spreadsheet, the GIS or the external relational spatial database.

Compartment models described by ordinary differential equations can be solved with macros, which implement algorithms for numerical calculation (Eulcr, Runge-Kutta). In spite of a lower efficiency of calculation, a wide use of the spreadsheet programs, built-in calculation functions, available graphs and a standard programming interface minimize training demands. A number of data formats enable data exchange between the spreadsheet and the GIS. A basic schema of the dynamic model in the spreadsheet environment illustrates Figure 13. Visual Basic macros can carry out all the calculations.

The worksheet contains a list of differential equations supported by algebraic equations and time series of input variables. The calculated outputs are written into the time series of output variables. Simulation parameters list the setting of numerical methods [27].
In case of larger models or using of special numerical algorithms, some mathematical programs for numerical calculation of differential equations assist as the model solver. Simulation tools like for example Advanced Continuous Simulation Language (ACSL), MATLAB-Simulink or Mathematica mostly can share external data through ASCII or XML formats [28].

3.6 ArcGIS Developments Tools

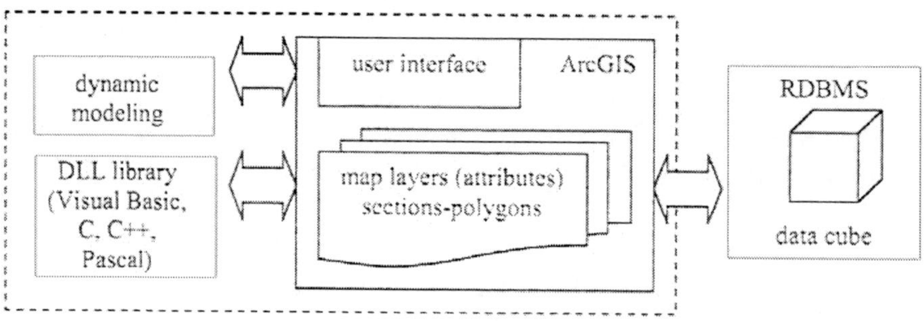

Figure 14. A basic schema of dynamic modeling implemented in the DLL library.

A number of present GISs is extended by programming kits, which offer development of software modules in a higher programming language. As using of GIS has become popular in water management, geospatial information has become available, including GIS map layers for stream network, watersheds and monitoring station locations. So, integration of the dynamic model into the ArcGIS environment offers dynamic modeling and spatial analysis in the frame of a common user interface with minimizing of an external data exchange. In addition to the standard programming tool, which is represented by Visual Basic extended by ArcObjects, the dynamic linked library (DLL) can be used to carry out simulation of the nitrate pollution. All the numeric algorithms are included in the DLL file that is called from the user interface of ArcGIS. A basic schema of dynamic modeling implemented in the DLL library illustrates figure 14. The ArcGIS user interface is extended by simulation tools, which enable setting of simulation parameters (numerical method, time step, simulation period) and running of the calculation. The initial and actual state values of variables are stored as attributes of the map layer that includes polygons of sections. Discrete time values of state variables in a specified time interval can be stored as copies of the original map layer. The schema in Figure 14 also illustrates the connection to a relational database, which can be in the frame of ArcGIS a personal database or an external database connected through the Spatial Database Engine (SDE). A symbol of the data cube in the relational database represents the three-dimensional data box for storing of data. Time series information is depicted in 3D space. The three coordinate axes can represent space (identified by an object code), variable being measured or calculated and discrete time. The 3D arrangement of data can support data presentation and multidimensional data analysis [29].

3.7 Individual Programs Developed with GIS Programming Libraries

As an example of modeling extended by the GIS spatial data, the small water basin of the stream Rakovnicky (the west part of Bohemia, the whole area of 368 km^2) is used to demonstrate simulation of nitrate accumulation and transport. The basin is divided into 9 main sections, which are traversed by the stream Rakovnicky. The reminder parts represent subbasins of subsidiary streams. The upper streams and down streams are surrounded by forest. The middle parts are mostly covered by agricultural areas, Figure 15.

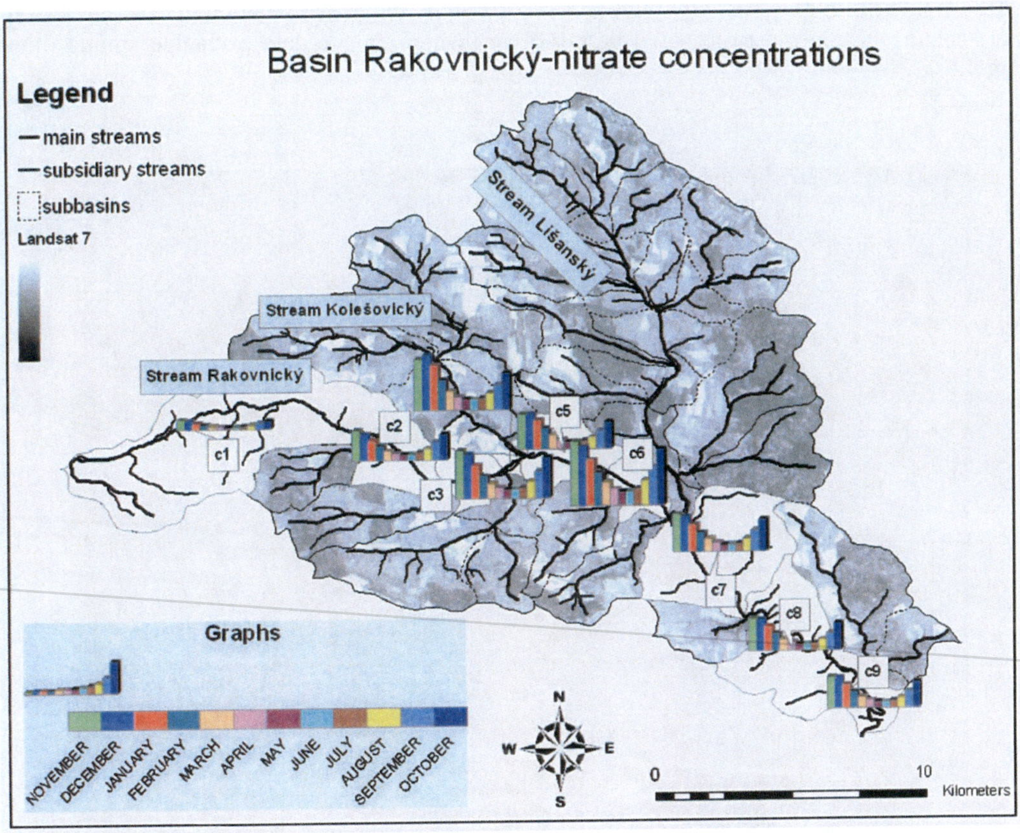

Figure 15. A map schema of the basin Rakovnicky implemented in a program for simulation of nitrate pollution with a dynamic model extended by spatial data [27].

The compartment model is formed by 15 ordinary differential equations that describe the dynamics of nitrates and consequently changes of the concentrations c_{11}, ..., c_{14}; C_{11}, ..., C_{13};

c_2, ..., c_9. The first section is subdivided into 7 parts to simulate nitrate interactions in substreams c_{11}, ..., c_{14} and reservoirs C_{11}, ..., C_{13}. The resulting concentration c_i is the average of all the concentrations calculated in the first section (c_{11}, ..., c_{14}; C_{11}, ..., C_{13}). Considering to available data, which are obtained from state authorities, mapping agencies and research projects, the estimates of model parameters and simulation are carried out for a period of one year (November 1999-0ctober 2000) in a daily time step. Meteorological and hydrological data together with information from the satellite Landsat 7 TM and ASTER are used to adjust model parameters.

As a case study, the standalone software application has been developed to test calculation of the described dynamic model. GIS functionality is provided by MapObjects development tools in the environment of Microsoft Visual Studio, Figure 16. The colored areas mark a level of nitrate concentrations in each section of the stream RakovnickY. On the left side, a series of map schemas illustrates the levels of concentrations in selected time points. On the right side, the more detailed map schema enables to display other map objects together with the digital maps and the satellite or aerial images. In addition to saving of time series of concentration in special binary or text formats, the development tools can manage to store spatial data in most of spatial data formats, which enable data exchange among other GIS.

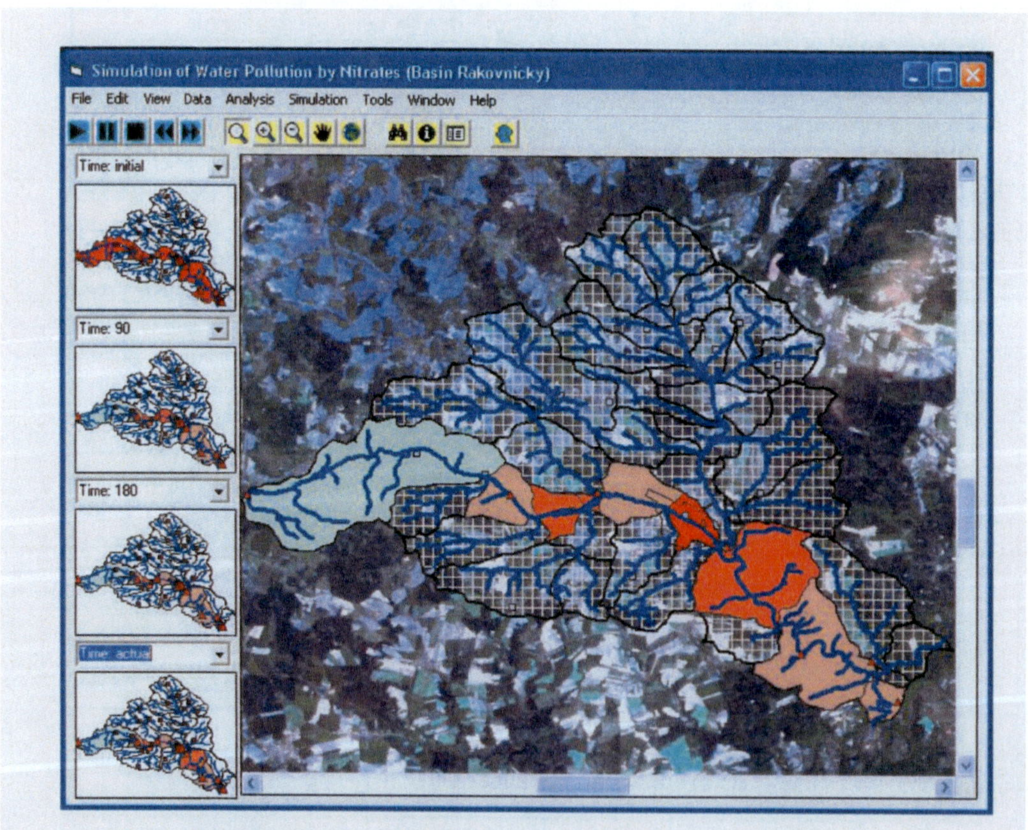

Figure 16. An example of the program for simulation in the environment of GIS tools.

Simulation of nitrate distribution with the introduced program has been provided on a few levels. The first level represents setting of parameters to equalize model outputs and data obtained from terrain measurements. Other levels cover simulation of various scenarios, which means for example reduction of point and non-point sources of pollution in a case of changes in agricultural and industrial management, pollution of the streams in case of environmental disaster, etc. Using of modeling extended by GIS data enables more complex analysis of observed processes. Classification of water, agricultural, forest and urban areas is provided with the data from Landsat 7 TM and ASTER. Whereas the agricultural and urban areas mostly act as sources of nitrate pollution, forest zones operate as sinks. Considering to the conditions for denitrification, high amounts of nitrates are intercepted outside of agricultural and industrial areas mostly in summer months. In case of basin RakovnickY, main sources of nitrate pollution exist in the median stream section. The downstream section covered by forest indicates the decreasing trends in the nitrate concentrations. According to a lack of terrain measurements and a number of available satellite scenes, the dynamic model based on proposed hypothesis is used to compensate many of these imperfections. In case of nitrate pollution in the basin Rakovnicry, dynamic modeling extended by spatial interaction in the environment of GIS enables forecast of nitrate concentrations in each section. Simulation of various environmental management scenarios improves decision-making processes in the basin

4. CONCLUSIONS

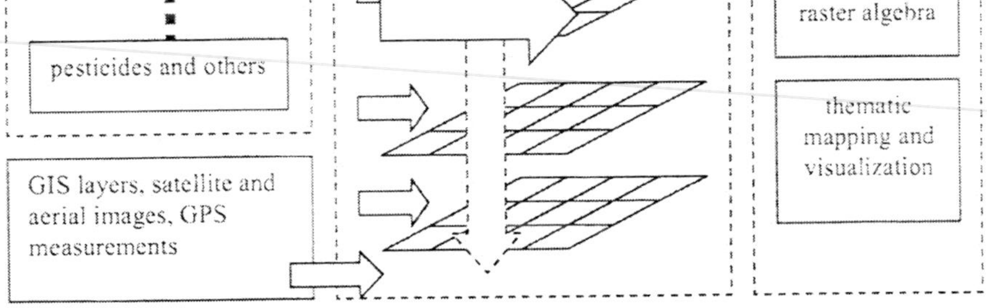

Figure 4. The basic structure of the compartmental model for modeling of pollutant transport [8].

Likewise in the example of a standalone software application for modeling of nitrate pollution in the basin, modeling of pollution by heavy metals, organic compounds, total phosphorus or pesticides can be developed in a similar way. Despite different interactions of a selected pollutant in an ecosystem, integration of dynamic modeling together with GIS can be developed by using of similar methods. Compartment models are able to provide basic descriptions of studied processes and GIS can manage all the spatial data, time series and data exchange. In case of pollution of a river basin by a few pollutants, more complex analysis can be carried out by integration of various described strategies in the frame of GIS, which is illustrated in figure 17.

Spatially distributed data focused on water pollution in the river basin play an important role in environmental protection. Only a complex assessment brings effective management of the whole basin and consecutively can support environmental planning and policy making. Dynamic modeling together with satellite data in the frame of GIS research represents new ways to solve all the problems.

ACKNOWLEDGMENTS

The paper covers research results created in the frame of a few projects. Various strategies of integration of dynamic model ing and GIS were tested in the frame of the project Environmental modeling in GIS supported by The Royal Society in London. The experimental data and digital maps in the individual examples have been collected and licensed in the frame of projects supported by Charles University in Prague. The software applications, satellite data and thematic maps have been processed in the GIS Laboratory at the Faculty of Natural Science, Charles University in Prague.

REFERENCES

[1] Fedra, K. (1996). Distributed Models and Embcdded GIS: Integration Strategies and Case Studies. In M.F. Goodchild. L.T. Steyaert. B.O. Parks (Ed.), GIS and Environmental Modeling: Progress and Research Issues (pp. 413-417). Fort Collins: GIS World, Inc.

[2] Merwade, V., and Maidment, D. (2002). Hydrologic Modeling. In: D. Maidment (Ed.), ArcHydro: GIS for Water Resources (pp. 167-175). Redlands, California: ESRI Press.

[3] Falconer, R.A. (1991). Review of Modeling Flow and Pollutant Transport Processes in Hydraulic Basins. In L.c. Wrobel and C.A. Brebbia (Ed.), Water Pollution: Modeling, Measuring and Prediction (pp. 3-23). London: Elsevier.

[4] Parker, P., Letcher, R., Jakeman, A., Beck, M.B., Harris, G., Argent, R.M., Hare, M., Pahl-Wostl, c., Voinov, A., Janssen, M., Sullivan, P., Scoccimarro, M., Friend, A., Sonnenshein, M., Baker, D., Matejicek, L., Odulaja, D., Deadman, P., Lim, K., Larocque, G., Tarikhi, P., Fletcher, c., Put, A.. Maxwell, T., Charles, A., Breeze, H., Nakatani, N., Mudgal, S., Naito, W., Osidele, 0., Eriksson, I., Kautsky, U., Kautsky, E., Naeslund, B., Kumblad, L., Park, R., Maltagliati, S., Girardin, P., Rizzoli, A., Mauriello, D., Hoch, R., Pelletier. D., Reilly, J., Olafsdottir, R. and Bin, S. (2002). Progress in integrated assessment and modeling. Environmental Modeling and Software 17: 209-217.

[5] Matejicek, L. (2003). Development of Software Tools for Ecological Field Studies Using ArcPad [CDROM]. In 2003 ESRI International User Conference Proceedings, San Diego, California, July 7-13. Redlands, California: ESRI.

[6] Jørgensen, S.E., and SOrensen, B.H., Nielsen, S. . (1996). Handbook of Environmental and Ecological Modeling. Boca Raton, Florida: CRC Press.
[7] Burke, R. (2003). Gelling to Know ArcObjects: Programming ArcGIS with VBA. Redlands, California: ESRl Prcss.
[8] Zeiler, M. (1999). Modeling Our World- The ESRI Guide to Geodatabase Design. Redlands, ESRI Press.
[9] Mitchell, A. (1999). The ESRJ Guide to GIS Analysis: Volume 1: Geographic Pallerns and Relationships. Redlands, California: ESRI Press.
[10] Stefanov, W.L., Ramsey, M.S. and Christensen, P.R. (2001). Monitoring urban land cover changes: An expert system approach to land cover classification of semiarid to arid urban centers. Remote Sensing of Environment 77, 173-185.
[11] Schmugge, T., French, A., Ritchie, J.C, Rango, A. and Pelgrum, H. (2002). Temperature and emissivity sepatation from multispectral thermal infrared observations. Remote Sensing of Environment 79: 189-198.
[12] Zhu, G., Blumberg, D.G. (2002). Classification using ASTER data and SVM algorithms; The case study of Beer Sheva, Israel. Remote Sensing of Environment 80, 233-240.
[13] Yamaguchi, Y., Fujisada, M., Kudoh, M., Kawakami, T., Tsu, H., Kahle, A.B., Pniel, M. (1999). ASTER instrument characterization and operation scenario. Adv. Space Res., 23 (8), 1415- I424.
[14] Puijenbroek, PJ.T.M., Janse, J.H. and Knoop, J.M. (2004). Integrated modeling for nutrient loading and ecology of lakes in The Netherlands. Ecological Modeling 174, 127-141.
[15] Wade, AJ., Whitehead, P.G., Jarvie, H.P., Neal, C, Prior, H. and Johnes, PJ. (2004). Nutrient monitoring, simulation and management within a major lowland UK river system: the Kennel. Mathematics and Computer in Simulation 64, 307 -317.
[16] Leon, L.F., Booty, W.G., Bowen, G.S. and Lam, D.CL. (2004). Validation of an agricultural non-point source model in a watershed in southern Ontario. Agricultural Water Management 65,59-75.
[17] Grewal, M.S., Weill, L.R. and Andrews, A.P. (2001). Global Positioning Systems, Internal Navigation, and Integration. New Yark: John Wiley and Sons.
[18] Paine, D.P. and Kiser, J.D. (2003). Aerial photography and image interpretation. New Jersey: John Wiley and Sons.
[19] Olivera, F., Furnans, J., Maidment, D., Djokic, D. and Ye, Z. (2002). Drainage systems. In: D. Maidrnent (Ed.), ArcHydro' GIS for Water Resources (pp. 167-175). Redlands, California: ESRI Press.
[20] Matejicek, L., Benesova, L. and Tonika, J. (2003). Ecological modeling of nitrate pollution in small river basins by spreadsheets and GIS. Ecological Modeling 170: 245-263.
[21] Norton, M.M. and Fisher, T.R. (2000). The effects of forest on stream water quality in two coastal plain watersheds of the Chesapeake Bay. Ecological Engineering 14, 337-362.
[22] Hannon, B. and Ruth, M. (1994). Dynamic modeling. New York: Springer-Verlag.
[23] Moore, J.D. (1996). Hydrologic modeling and GIS. In M.F. Goodchild, L.T. Steyaert, B.O. Parks (Ed.), GIS and Environmental Modeling: Progress and Research Issues (pp.143-148). Fort Collins: GIS World, Inc.

[24] Arheimer, B. and Brandt, M. (2000). Watershed modelling of non-point nitrogen losses from arable land to the Swedish coast in 1985 and 1994. Ecological Engineering 14, 389-404.

[25] Prakash B., Teeter, L.D., Lockaby, B.G. and Flynn, K.M. (2000). The use of remote sensing and GIS in watershed level analyses of non-point source pollution problems. Forest Ecology and Management 128,65-73.

[26] Benesova, L., Matejicek, L., Matouskova, M. and Tonika, J. (2001). Ekologicke hodnoceni kvality Yody v malych povodich: Rakovnicky potok (Ecological assessment of water quality in small water basins: the stream Rakovnicky) [CDROM]. Final Report, Charles University in Prague.

[27] Matejicek, L. (2002). Ecological Modelling Using the Spreadsheet Model TabSim. In A.E. Rizzoli, AJ. Jakeman (Ed.), Integrated Assessment and Decision Support, ist Biennial Meeting of the international Environmental Modellling and Software Society (pp. 354-359). Lugano: iEMSs.

[28] Matejicek, L. and Tlapakova, L. (2003). Modeling in the GIS environment: dynamic processes of water pollution. Acta Universitatis Carolinae Environmentalica 13,47-62.

[29] Legendre, P. and Legendre, L. (1998). Numerical ecology. Amsterdam: Elsevier.

In: Environmental Modeling with GIS
Editor: Lubos Matejicek, pp. 55-70

ISBN: 978-1-60876-363-4

Chapter 4

SPATIO-TEMPORAL MODELING OF THE DUST EMISSIONS FROM AN OPENCAST COAL MINING AREA

Lubos Matejicek
Institute for Environmental Studies, Charles University in Prague, Faculty of Natural Science, Benatska 2, 128 01, Prague, Czech Republic
Zbynek Janour
Institute of Thermomechanics, Academy of Sciences
Dolejskova 5, 182 00, Prague, Czech Republic

ABSTRACT

The described research is carried out to map the dust emission sources of opencast coal mining activities and their influence on surrounding areas. The results can be utilized by mine authorities and environmental scientists in the field of soil pollution and air quality monitoring for advance prediction.

The attached case study is focused on the opencast coal mine area in northwest Czech Republic. To achieve the objectives, spatial data from remote sensing represented by images from Landsat 7 are used for identification of large opencast coal mine areas. Data from GPS and laser scanning complemented by geodetic measurements serve as an input for building of the digital elevation model. Data inputs based on aerial images and thematic maps include spatial locations of the local dust emission sources represented by transport routes, temporary repositories and mining excavators. Spatial estimates of the dust emission rates are validated by conducting a field study. All the spatio-temporal data are integrated in the framework of the GIS project that contains complex data structures supporting advanced spatial analysis and visualization.

The influence of dust emissions on surrounding areas is estimated by dispersion modeling based on U.S. EPA models. The standalone simulation tools managed by ISC-AEROMOD View are utilized to assess deposition fluxes from selected emission sources. After the post-processing stage, simulation outputs are incorporated into the GIS project.

All the spatial data together with the dispersion modeling outputs are displayed in GIS as the map layers. Due to the obtained results of the impacts on residential areas in the neighborhood of the opencast coal mine, remediation procedures focused on relocation and reduction of the main emission sources are needed to minimize the influence of the coal dust on living environment. In this case, GIS can assist to optimize the opencast coal mine infrastructure considering to the decreasing major mining activities that cause the significant dust emissions and soil pollution.

Keywords: Opencast coal mine; emission sources; GIS; dispersion modeling.

1. INTRODUCTION

Dust emissions originated from open sources in the coal mine areas are mainly emitted at a few stages that include mine excavators, transport routes, temporary repositories and dispatch of coal from the mine. Concentrations of the dust released from opencast mining activities vary greatly from one area to another depending on the intensity of local sources, terrain profiles, general weather conditions, and surface soil moisture. The extensive use of coal as an energy source and other attendant pollution effects is caused by industrial proliferation. The impact that this pollution had on health was profound, which led to the emission control measures in many industrial areas with attendant improvement in health indices. Dust can cause ill health by skin contact, eye contact and inhalation. Besides individual sensitivity, the real risk of injury depends on duration and level of exposure. It affects not only the employees working in the plant but also the community in the neighborhoods.

Models used for estimating of atmospheric dust concentrations and deposition impacts are mostly based on air dispersion modeling approved by local agencies and government authorities. They incorporate transport, dispersion and deposition of pollutants in the atmosphere using input data for particulate matter and air-flow conditions. The numerical models are often represented by standalone computer programs that share data throughout the files by simple pre-processor linkage and post-processor linkage. Considering to spatial distribution of the explored processes, more complex processing tools are needed in the framework of spatio-temporal modeling. Thus, the geographic information systems (GISs) are used as tools to capture, modify, process and display spatio-temporal data. GIS can manage both geometry data (location, shape, and topology) and attributes that contain the properties of spatial objects (emission rates, concentrations, identification and description).

In case of spatio-temporal modeling, two concepts of modeling are merged. In spatial modeling, the basic concept is one of location, of spatial distribution and spatial relationships. The basic concept in temporal modeling is one of state, determined in terms of numbers, mass or energy. The overlap of both concepts is apparent, which promotes integration of spatial analysis and numeric modeling in the framework of GIS. The resultant projects bring together spatial data (thematic soil mapping, digital elevation models, satellite images and aerial images, emission inventories, meteorological data, receptor locations for measured and modeling impacts) and simulation outputs (temporal concentrations of pollutants for short episodes and long-term frequency data, emissions and impact scenarios for optimization modules). Many aspects and examples of spatio-temporal modeling in the framework of GIS are discussed in the book "GIS and Environmental Modeling: Progress and Research Issues" edited by Goodchild [1]. Nowadays, the relationship of spatial aspects and temporal aspects is going to be reconsidered with the increasing power of computer hardware and GIS software. Thus, in a number of simulation tasks, there is no need to couple GIS with the special software designed for high performance in dynamic modeling.

2. Emission Sources

The dust emissions from an opencast mining area come from a wide variety of sources. The main objective of the present study is to assess the impact of the predominant dust emission sources originated from mining activities and coal mine surface. The dust emissions are emitted directly into the atmosphere and in the framework of subsequent dispersion modeling represent primary pollutants. For primary pollutants, a proportional relationship can be estimated between emissions and ambient concentrations. However, for a secondary pollutant, reducing emissions of the precursor have not necessarily lead to a proportional reduction in its ambient concentration. Estimates on the rate of emission from different sources are vital for the development of abatement strategies. By identifying the major sources, the best means of setting the dispersion models and improving dust deposition in the neighborhood can be resolved. In addition to the source size, the amounts of the dust released from coal mine sites depend on the level of mining activities (coal extraction, transport frequency, coal sorting), surface soil state (soil type, soil humidity and temperature) and meteorological conditions (wind speed, wind direction, rainfall, air temperature and air humidity). Information on emission conditions and knowledge of emission processes are sufficient to carry out a good estimate of emissions from many sources. But in many cases, emissions depend on local conditions, as well as on pollution abatement technology. Thus, emission inventories for such emission sources are therefore liable to be less accurate.

Dust differs from a single pollutant. It is made up of particles of many different sizes and chemical composition. Generally, the concentration of airborne particles can be measured in several different ways. Formerly, estimates have tended to include total suspended particles and/or black smoke, but increasingly estimates of PM_{10} emissions (particles less than 10 μm) are being made. Considering to the different dominant emission sources in the US, Europe and other parts of the world, the estimated contributions of the main sources to total PM_{10} emissions vary. In the US, the inventory for PM_{10} indicates that the vast part of emissions originated from roads, agriculture or forestry. In Europe, the inventories suggest that the dominant source of emissions is fossil fuel combustion in industry zones or urban areas and road transport. Based on the US EPA reports, fugitive dust emissions and wind erosion together often represent the majority of the total US emissions. But these emissions are very sensitive to regional soil conditions and weather. Thus, the direct implementation of US emission estimates to Europe is difficult. Moreover, airborne PM_{10} is used to be divided into fine and coarse PM_{10}. Fine particles are often considered as the $PM_{2.5}$ fraction (particles less than 2.5 μm) and coarse particles form the reminder. The fine particles originated mainly from fossil fuel combustion, vegetation burning, metallurgical industry and secondary PM, which are formed in the air based on chemical reactions with other pollutants and atmospheric gases.

The estimates of atmospheric emission are based on major mining activities by considering type of mining, transport technology, methods of sorting and other redistributions. The data used to estimate emissions must be constantly enhanced by actual measurements of sources related to the emissions. These are usually the size for stationary places, such as mining areas and sorting places, and traveled distance for mobile sources, such as coal conveyors and trains. Emissions factors based on individual sources are then multiplied by the appropriate statistics, which can be mainly derived from intensity of mining, surface soil humidity and wind speed. Generally, emission inventories contain data for three shape categories of emission sources: point sources (small mining areas, sorting places, excavators and redistribution sites), line sources (roads and railways) and area sources (larger mining areas and bare surface of opencast coal mines). The detail descriptions of the emission sources and their inventories for individual pollutants are included in the US EPA reports accessible via official web pages (www.epa.gov). The impacts of air pollution on human health, methods for pollutant monitoring and identification of potential emissions are described by many authors in the books, papers and research reports. For example, the multidisciplinary book "Air pollution and health" edited by Holgate [2] brings together a number of issues relating to the monitoring of air pollution and description of UK's National Atmospheric Emission Inventory (NAEI). Other reports and databases of emission factors for the main industrial and transport sources, which assist in compiling emission inventories for British cities, are also available from web pages. In case of coal mining industry, local agencies and authorities are responsible for monitoring of dust emissions and other air pollutants in the framework of each national territory [3, 4].

3. Remote Sensing, GPS And GIS

Potential emission sources can be identified by remote sensing methods in dependence on the spatial scale from a number of image data sources. Remote sensing gathers information at altitude from the energy fields arising from the Earth's surface to explore their spectral, spatial and temporal properties. Nowadays, a number of remote sensing systems can assist in gathering information that is focused on the surface of the opencast mining areas. The system by itself is divided into three basic parts: the scene, the sensor and the processing system. The scene consists of the Earth's surface and the intervening atmosphere. In case of passive optical systems, the portion of the spectrum relies on solar illumination. The sensor collects the main body of the data to be used for spatial analysis in the processing system. Both the sensor system and the processing system are under human control to provide requisite image data and spatial analysis [5].

The larger mining areas can be detected from satellite images, which offer multispectral data. For example, one of the longest-running earth-observation satellite systems is represented by the Landsat satellites. Since 1972, six satellites have been successfully lunched. The latest satellite, Landsat 7 (lunched in 1999), carries the Enhanced Thematic Mapper (ETM+) sensor, which provides 7 bands of multispectral data at 30 meters resolution for bands 1 (0.450-0.515 μm, blue), 2 (0.525-0.605 μm, green), 3 (0.630-0.690 μm, red), 4 (0.775-0.900 μm, near IR), 5 (1.550-1.750 μm, mid IR), 7 (2.090-2.350 μm, mid IR) and at 60 meters resolution for the band 6 (10.40-12.50 μm, thermal) [6]. The panchromatic band

(0.520-0.900) at 15 meters resolution can be used to add crispness and detail to the multispectral data. The swath width of 185 km provides an adequate size of the surface area for classification and identification of the dust emission sources from individual images. The bands 3-2-1 (red, green, blue) create a true color composite that displays objects similar to a color photograph. The bands 4-3-2 (near IR, red, green) and bands 5-4-2 (mid IR, near IR, green) create a false color composite or a pseudo color composite, which means that object colors do not reflect the feature in natural colors. To identify potential dust emission sources, the band 7 (mid IR) is important for discrimination of geologic rock types, soil boundaries and soil moisture content [7]. In case of the following study, the Landsat 7 image is used for identification of potential area emission sources on the opencast coal mine surface.

The smaller emission sources (small mining areas, local coal repositories, redistribution places, sorting places, roads and railways) can be identified from high resolution satellite images or from aerial photographs. In case of the aerial photographs, the aerial spatial data at 0.5 meter resolution in the visible part of the spectrum are used for mapping of the local mining activities. The aerial images are used together with thematic map layers in GIS for identification of the local emission sources.

The range of possible degrees of human/machine participation in the image analysis process can be divided into a few steps: fully manual methods, machine aided manual methods, manually aided machine methods, and automatic analysis [5]. Whereas identification of the large emission sources from Landsat 7 can be provided by manually machine methods or automatic analysis, specification of the local mining activities requires manual methods or machine aided manual methods extended by terrain mapping.

In case of active systems (radar, lidar), the illumination mechanism is incorporated into the sensor system. Among a wide range of actual remote sensing techniques, the 3D laser scanning is used to collect data for the digital elevation model (DEM) of selected mining areas. The laser (Light Amplification by Stimulated Emission of Radiation) consists of photons of light of a single wavelengths aligned in the same direction in a narrow beam. Using of laser technology in the framework of remote sensing is often declared as LiDAR (Light Detection And Ranging), which involves aiming a light beam at a target and having it bounce back to the source. The position of target points on the surface is calculated from the time measurement of the beam flight and direction the beam was aimed in. By aiming millions of laser beams at the selected part of the surface, the DEM can be created for deriving map layers based on contour lines, slopes and aspect ratio.

After saving data from remote sensing systems, spectral corrections and spatial processing are needed to make images more interpretable for a particular application. In addition to optimization of contrast and brightness, spectral enhancement based on transforming the values of each pixel on a multiband basis is carried out. In the framework of spatial processing, georeferencing needs to be provided to assigning map coordinates to the image data. Georeferencing includes the rectification, which is the process of projecting the image data onto a plane and making it conform to a map projection system [7]. The projection coordinates are mostly collected by global positioning system (GPS).

GPS has been utilizing for non-military research for a few decades. In basic principle, a GPS receiver calculates its position by measuring the distance between itself and accessible GPS satellites. The receiver identifies each satellite's signal, which also contains information about the satellite's position and the precise time. Besides built in techniques to improve the positional accuracy and using the phase mode, the network of the surface reference stations in the Czech Republic is used to make positional corrections. Thus, the estimates can be realized in the scale of centimeters of the absolute positional accuracy.

Finally, a spatial modeling approach is desirable for integration of images from remote sensing, measurements from GPS, and spatial data from thematic mapping. Thus, the GIS software is used for data management, spatial analysis and visualization. In addition to storing and maintaining spatial data, the new map layers based on classification of satellite images and aerial images can be created in the framework of a GIS spatial database to support dispersion modeling and decision making processes. Despite an increased deal of time and effort to produce new data, GIS fits the character of emission science as well as the technical environment of the expected end users. In addition to identification of the large emission sources from satellite images and local emission sources derived from aerial images, mobile emissions caused by the coal transport and the secondary emitted dust along the roads represent other significant emission sources. They are mapped as the line emission sources in the GIS environment. Mobile emission rates vary by transport activities and surface conditions. The relationship between the emission rates and the surface location is also the primary argument for using GIS in spatio-temporal modeling of the dust emission rates. Thus, the mobile sources and stationary sources are included into the GIS spatial database together with other spatio-temporal data (thematic maps, terrain measurements and observations, input/output of dispersion modeling).

GIS activities in spatio-temporal modeling of emission rates are mostly focused on air quality modeling. A number of research efforts have taken advantage of the GIS spatial environment to perform the data inputs and to visualize the simulation outputs of the standalone numerical models. Maguire et al. [8] presents various applications concerned with modeling and spatial analysis within a GIS framework. The described techniques include various aspects of linking models and GIS software. Examples of environmental modeling document possibilities of the GIS software to solve spatio-temporal phenomena. A number of studies deal with characterization of urban air quality supported by GIS as a management system [9, 10]. For example in 2000, Bachman et al. [11] published a conceptual framework for a GIS-based emission model extended by the MEASURE model (Mobile Emission Assessment System for Urban and Regional Evaluation). The GIS-based model for public release is focused on regional mobile source emissions. Despite the pollutant types based on VOCs, NO_x and CO, the concepts of innovative use of GIS and spatial analysis can be partly implemented into the spatio-temporal modeling of dust emissions. Many modeling concepts, description of modeling tools and air quality measurement reports are published by US EPA.

The implementation of spatio-temporal concepts depends on target software tools. In case of ArcGIS, the advanced techniques based on the ESRI's geodatabase and development tools (ESRI's ArcObjects) can be used to manage data and to build up new analysis functions. The used data formats and a wide range of export/import functions guarantee possibilities of sharing data together with other software tools. Generally, coupling of standalone software systems with GISs can be based on three types. The low level coupling is represented by exchanging files. In this case, the GIS program is mostly used to prepare the data inputs, which are then passed to another software package, and after execution, the results are returned to the GIS tools for display and spatial analysis together with other map layers. The higher level includes sharing programming standards such as Microsoft .NET or Java that enable common running of functions and methods through single scripts. The final step can be represented by an object oriented programming application based on software development tools and programming libraries that cover required functionality. In case of the higher level coupling, the software tools can run more efficiently, but more experience is needed to integrate all the software modules.

4. Dispersion Modeling in the Framework of GIS

Modeling of atmospheric dispersion can be provided by a number of computer programs, which are available on the Internet or offered commercially together with user extensions. The Industrial Source Complex Short-Term model (ISCST3) is used to estimate atmospheric dispersion of the dust emissions from opencast mining areas. It provides options to model emissions from various sources at industrial zones or urban areas. The model is based on the straight-line, steady-state Gaussian plume equation. Emission sources are divided into four basic types: point sources, volume sources, area sources, and open pit sources. Other emission sources can be put together from the basic types. Thus, the volume source option and the area source option may be used to build up line sources.

The modeling tools accept hourly meteorological data records to define the conditions for plume rise, transport, diffusion and deposition. The concentration or deposition value is estimated for each source and receptor combination for each hour of input meteorology. As the outputs, user-selected short-term averages are calculated. The dry deposition flux, the wet deposition flux, or the total deposition flux can be estimated at each receptor location. The detail information about ISCST3 is included in the US EPA reports. Comparison of ISCST3 and other models, model evaluation, case studies for various pollutants and industrial areas are published in a number of papers [12, 13].

In case of running the ISCST3 model in the ISC-AERMOD View interface, the coupling of the air dispersion modeling system with GIS is based on exchanging files. The input data are represented by the meteorological data preprocessed by U.S. EPA PCRAMMET program, data focused on emission sources preprocessed by U.S. EPA Building Profile Input Program (BPIP), and simulation parameters. The output data are included in the simulation reports, which can be imported into GIS or handled by the post-processing utility POST View to display simulation results.

5. Visualization of Spatio-Temporal Data

Two concepts of spatio-temporal modeling require more complex visualization tools. In spite of that each computer program for simulation of temporal phenomena contains special display tools, the visualization system based on GIS principles is necessary to make realistic scenes that allow navigation and interaction with spatio-temporal data. In addition to display temporal data together with map layers, the unified storage formats, spatial measurements and statistical analysis are mostly included into GIS visualization extensions (ArcScene, ArcGlobe, ArcGIS-3DAnalyst) [14].

6. A Case Study of the Selected Opencast Mining Area

The following example illustrates spatio-temporal modeling of the dust emission rate from an opencast coal mining area. The example deals with data from remote sensing, GPS and dispersion modeling. In this case, the GISs are mainly used to manage complex data structures and to visualize simulation results from dispersion modeling. The data management is provided in the ArcGIS geodatabase, which supports all collected spatio-temporal data in various data formats (ARC/INFO's coverages, ArcView's shapefiles, CAD files, vector data, raster data, ASCII files). The ArcGIS geodatabase models are implemented on standard relational databases that manage data on a variety of different database platforms. The geodatabase model supports a variety of geographic object types with properties, behavior and relationships [15]. Thus, the geodatabase with GIS functionality forms a highly integrated environmental decision support system that can analyze emission rates and dust deposition together with spatial data in the map layers focused on demographic, social and economic phenomena.

6.1. Mapping of the Dust Emission Sources

The study site is located in northwest Czech Republic and covers a few square kilometers. The approximate view in Figure 1, captured from Google Earth, shows the surface of the opencast coal mine and its surrounding areas. The area of interest, which significantly affects residential areas on the right side of the image, is in Figure 2. The complex remediation of this area is planned to improve living environment in the region. Thus, the mapping of the dust emissions is important for decision making processes in feature.

Figure 1. The coal mining areas with the marked area of interest (Google Earth, August, 2007).

Figure 2. The area of interest (mapy.cz, August, 2007, www.mapy.cz).

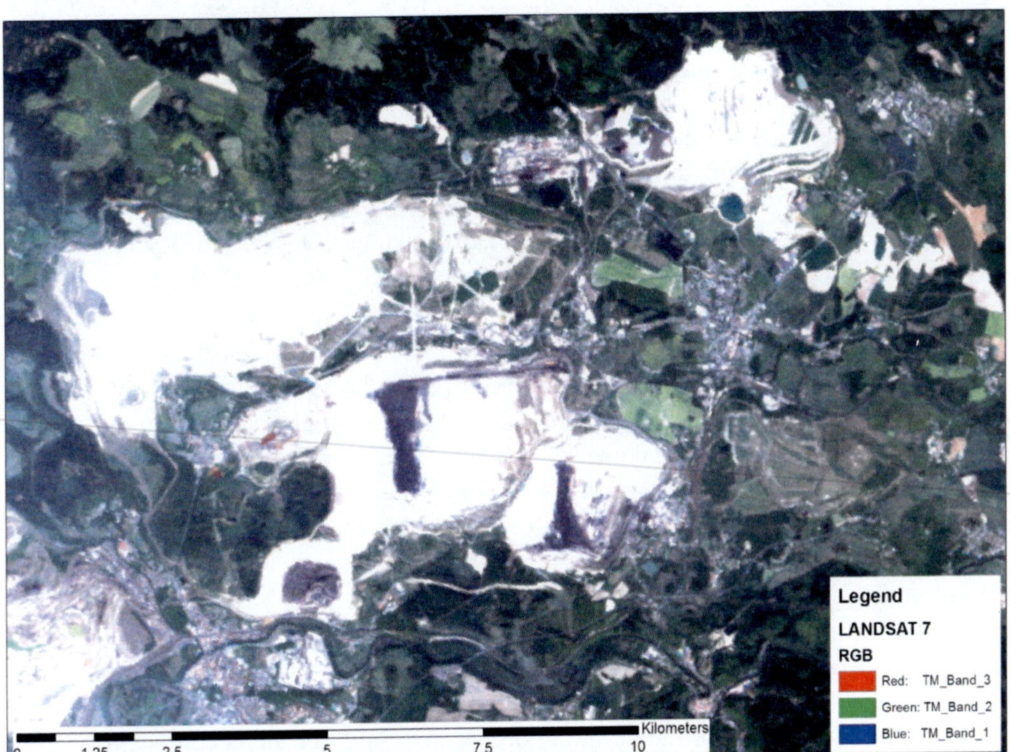

Figure 3. The Landsat 7 image created as a true color composite (bands: 3-2-1).

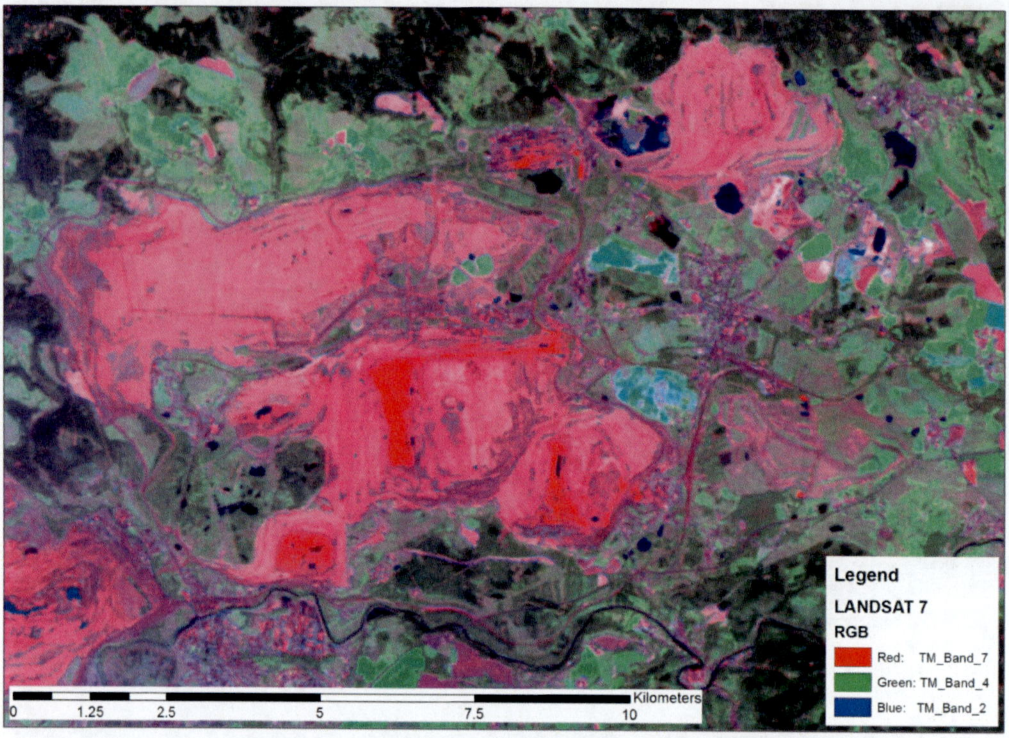

Figure 4. The Landsat 7 image created as a pseudo color composite (bands: 7-4-2).

Generally, the potential dust emission sources from large opencast mining areas are identified from multispectral satellite images. Figure 3 shows the Landsat 7 image that is created as a true color composite (bands: 3-2-1). Surface objects are displayed similar to their natural visible colors. The mining areas are light-colored. Figure 4 shows the Landsat 7 image represented by the middle infrared band 7 (2.090-2.350 μm), the near infrared band 4 (0.775-0.900 μm) and the visible band 2. The band 7 and the band 4 are transformed into the visible light spectrum. The band 7 (sensitive to soil boundaries and soil moisture) together with remaining bands indicate the potential emission sources by the light violet color.

As an example, the dust emission sources from the local mining activities in the sub-area of interest are identified from aerial images in Figure 2, and verified by terrain mapping. The individual repositories, the redistribution places, the sorting places, the roads and the railways can be directly recognized from the aerial image and included into the GIS spatial database together with attributes, which contain estimates of emission rates, identification and description.

The LiDAR system is used to scan a smaller part of the coal mine area containing the main coal repository that causes the predominant dust emissions in the neighborhood of the residential areas. The millions of points on the surface are used to create the precise DEM for a more detail identification of the dust emission sources, and for the conditions of the dust transfer by the wind circulation. The clouds of points for the DEM construction are in Figure 5. A part of the resulting DEM created by a triangulated irregular network is in Figure 6. Two spatial data sources for the DEM construction can be compared in the final scene. The middle of the DEM is created by the spatial data from LiDAR. The boundaries of the DEM originate from terrain measurements and contour lines. The scene shows a slope of a part of the opencast coal mine. The local railway, which is situated diagonally across the scene, is used for the coal transport to the local coal repository on the right side of the scene. After the sorting, the coal sorts are transported from the mine. Thus, this local area includes a higher density of various emission sources represented by the local repository, the railway corridors, the sorting places, the excavators and the redistribution sites. Furthermore, the area is located near the residential zones on the upper side of the slope, which requires continuous monitoring and regulations of mining activities. The projection coordinates for georeferencing of the aerial images and the clouds of points from LiDAR are provided by GPS. The major emission sources identified from aerial images and precisely located by GPS measurements are marked in Figure 6.

The spatial data from remote sensing and GPS measurements together with mapping of the emission sources are included in the GIS project. The main GIS tool for data management and spatial relationship is represented by the personal geodatabase in the environment of ArcGIS. The schematic view of the geodatabase focused on the data structure of the previously estimated emission sources is in Figure 7. The emission sources are divided into three major classes: the area sources, the line sources and the point sources. The geodatabase contains spatial data extended by attributes for each class of the emission sources. An example of the geodatabase view in the ArcCatalog environment shows a few symbolic icons that represent individual layers with the spatio-temporal data containing the emission sources.

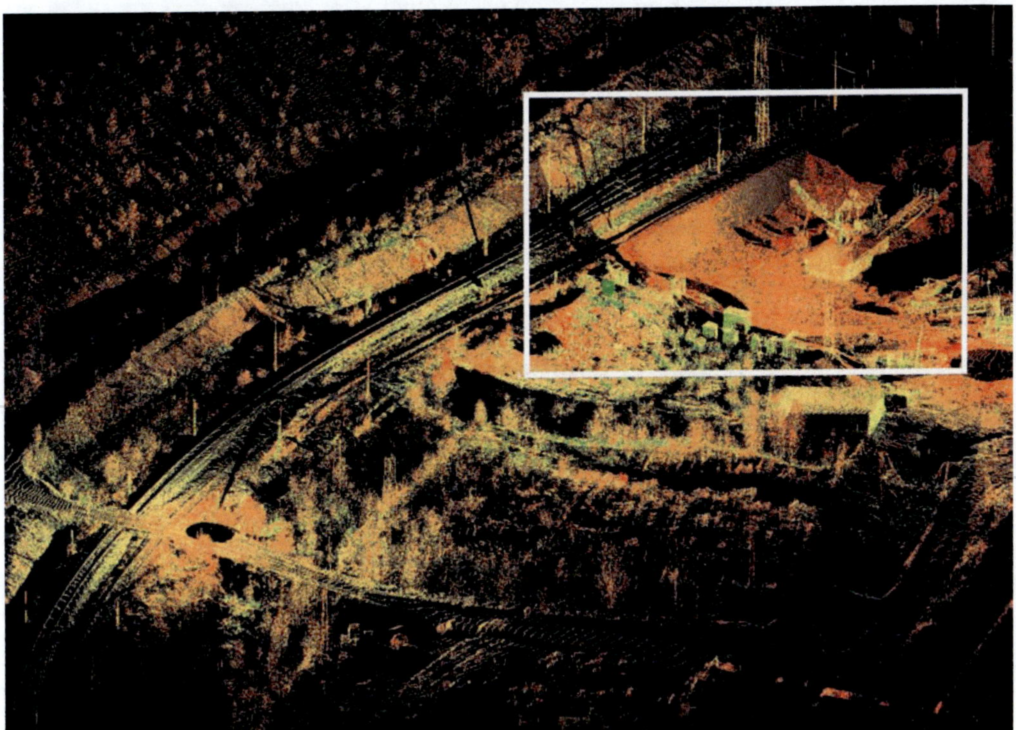

Figure 5. The clouds of points created by the LiDAR, the rectangle marks the area of interest.

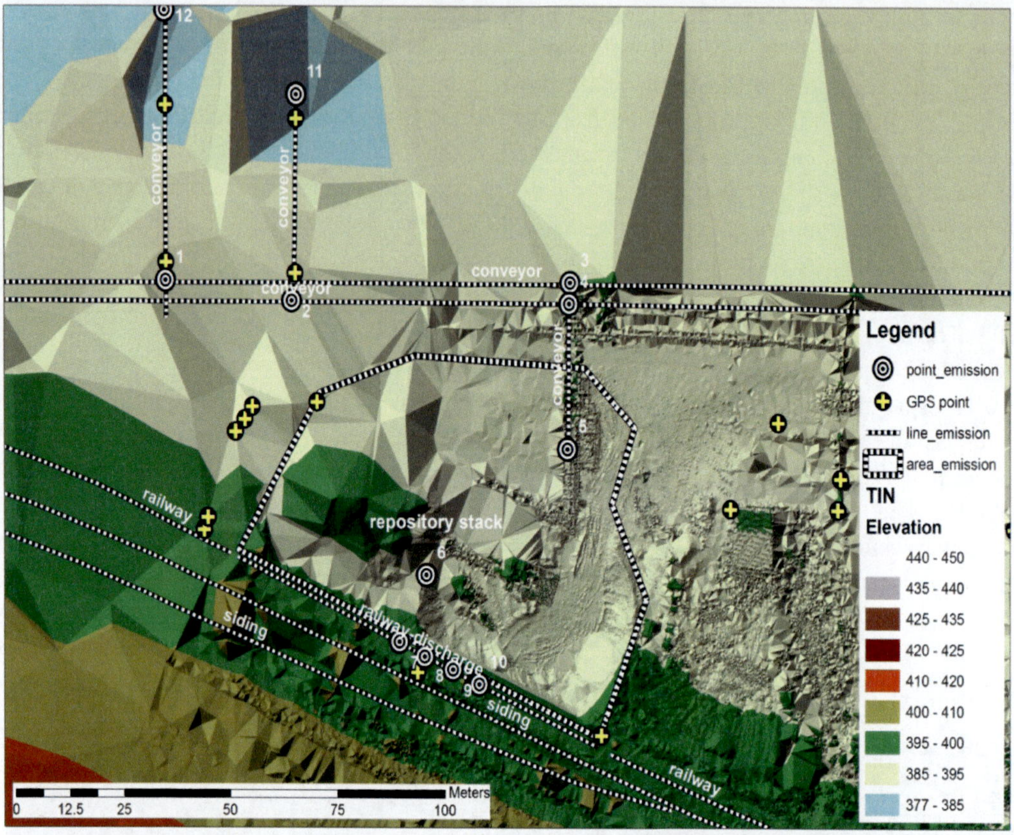

Figure 6. The area of interest with the potential emission sources complemented by the DEM.

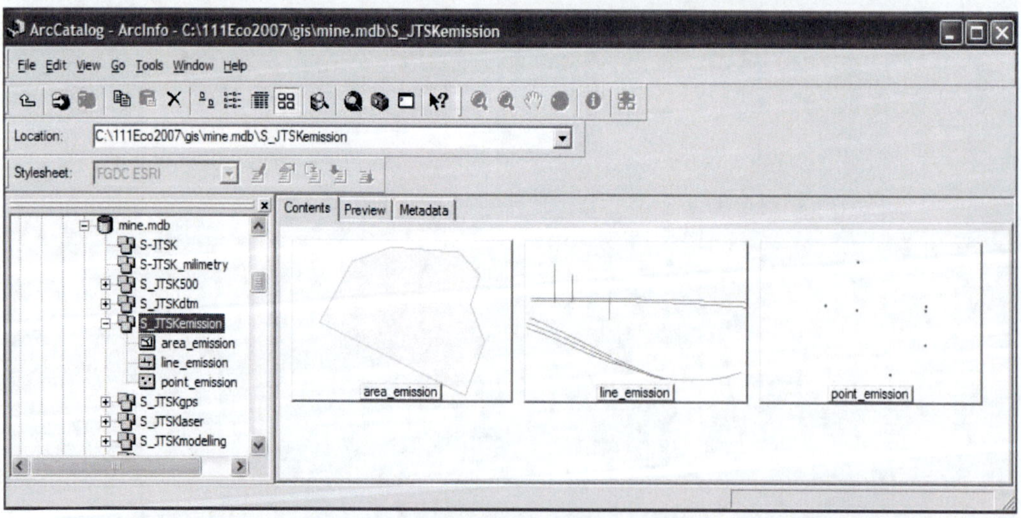

Figure 7. An example of the schematic view with the map layers focused on emission sources.

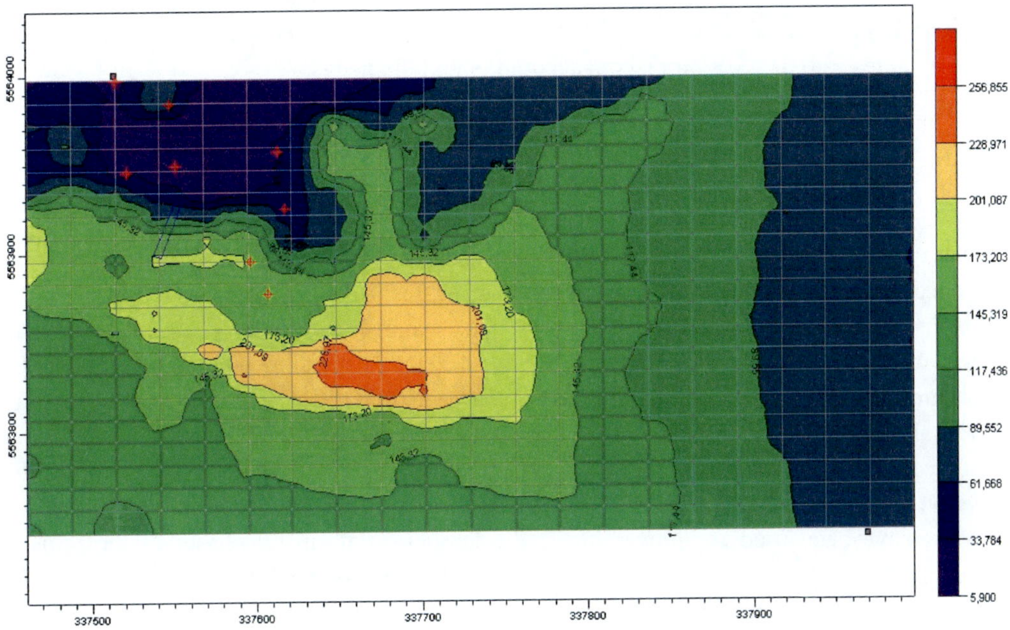

Figure 8. A schema of the dust concentrations originated from the local emission sources.

6.2. Sharing Data between Dispersion Modeling Tools and GIS

The used simulation software ISC-AERMOD View requires meteorological data in the specific format pre-processed by the U.S. EPA PCRAMMET and spatial data pre-processed by the U.S. EPA BPIP. The spatial data include emission sources, barriers for calculation of downwash effects, DEM and thematic map layers. The GIS is used for data management, transformation of coordinates between the local coordinate systems and the UTM, construction of the DEM from a few data sources (topographic maps, GPS measurements, laser scanning) and modification of data into other formats. The results of a case simulation for the local area of the opencast coal repository are in Figure 8. The contour plots produced by POST View in the environment of ISC-AEROMOD View show the dust concentrations originated from the selected emission sources marked on the upper left part of the local area. The results indicate significant variation in dust emissions in dependence on the wind speed, the wind direction and the DEM. The variation of emission rates from day to day caused by mining activities and soil humidity also affects the concentration levels. Thus, many case simulation studies are needed for proper setting of the dispersion model in the framework of various conditions.

6.3. Visualization of Spatio Temporal Data in GIS

As an example, the scene in Figure 9 shows levels of the dust concentrations near the surface. The layer with the surface concentrations is shifted to the upper part of the scene to enable visualization of the lower layer with the aerial image. Both the layers are draped on the DEM. Thus, the variations of the dust concentrations can be compared with the real objects on the aerial images together with the DEM. The scene can be extended by other map layers that are included into the GIS projects to accomplish all the needs.

7. CONCLUSION

The approach described in this contribution deals with mapping emission sources from the opencast coal mining area in the various spatial scales. As an example, a part of the mine area is used to demonstrate spatio-temporal data management in GIS and sharing data together with standalone programming tools focused on dispersion modeling. Finally, the spatio-temporal data integrated into the geodatabase of the GIS project are used for display and visualization. The case study also demonstrates how the DEM based on 3D laser scanning is applied in dispersion models and in final visualization. With the exception of the dispersion modeling based on ISCST3 in the framework of the ISC-AEROMOD View, the GIS tools have been used for processing spatio-temporal data in all phases of the project. Individual phases demonstrate processing of spatial data based on satellite images, aerial images, laser scanning, GPS, digital thematic maps and terrain observations. The temporal data are formed by meteorological conditions, variable activities of emission sources and soil humidity. Due to the high variation of temporal data, only the approximate estimates of emission rates of selected emission sources have been used for dispersion modeling to demonstrate visualization of simulation results in a more straightforward way.

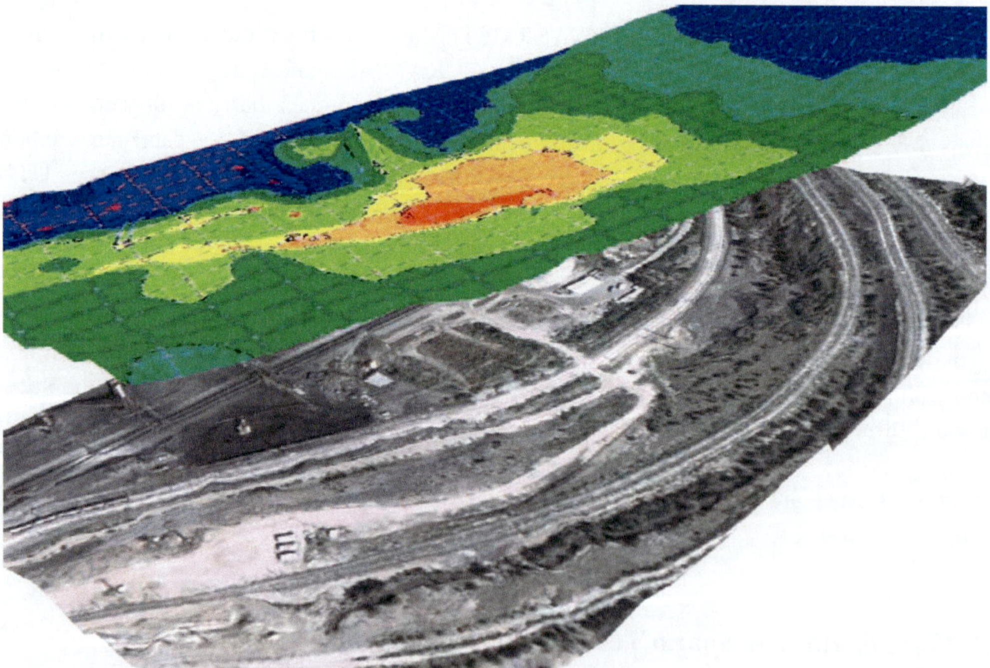

Figure 9. The scene with the layers containing the contour lines and the aerial image.

In this case, spatio-temporal modeling managed by GIS can support environmental planning and policy making. GIS by itself provides direct access to a large volume of data. Its functionality includes a wide range of analytical methods that can be used for interpretation, scenario analysis and forecasts. Thus, it can bring the best available knowledge to bear on decision making processes and reach a broad audience.

ACKNOWLEDGMENT

The presented research was carried out in the framework of the project supported by Academy of Sciences in the Czech Republic (AVCR 1ET400760405). The spatio-temporal data and terrain measurements have been processed by ArcGIS, ERDAS Imagine and ISC-AEROMOD View in the GIS Laboratory at the Faculty of Natural Science, Charles University in Prague.

REFERENCES

[1] Fedra, K. (1996). Distributed Models and Embedded GIS: Integration Strategies and Case Studies. In M.F. Goodchild, L.T. Steyaert, B.O. Parks (Ed.), *GIS and Environmental Modeling: Progress and Research Issues* (pp. 413-417). Fort Collins: GIS World, Inc.

[2] Holman, C. (1999). Sources of Air Pollution. In S. Holgate, H. Koren, J. Samet, R. Maynard (Ed.), *Air Pollution and Health* (pp. 115-148). Electronic book: Elsevier.

[3] Lu, H. and Shao, Y. (2001). Toward quantitative prediction of dust storms: an integrated wind erosion modelling system and its applications. *Environmental Modeling & Software 16*, 233-249.

[4] Chakraborty, M.K., Ahmad, M., Singh, R.S., Pal, D., Bandopadhyay, C. and Chaulya, S.K. (1999). Determination of the emission rate from various opencast mining operations. *Environmental Modeling & Software 17*, 467-480.

[5] Landgrebe, D.A. (2003). *Signal Theory Methods in Multispectral Remote Sensing*. New Jersey: John Wiley and Sons.

[6] Richards, J.A. and Jia, X. (2006). *Remote Sensing Digital Image Analysis*. Berlin: Springer.

[7] Stacey, S. and Pouncey, R. (1997). *Erdas Field Guide*. Atlanta: Erdas, Inc.

[8] Maguire, D.J., Batty, M., and Goodchild, M.F. (2005). *GIS, Spatial Analysis and Modeling*. Redlands, California: ESRI Press.

[9] Puliafito, E., Guevara, M., Puliafito, C. (2003). Characterization of urban air quality using GIS as a management system. *Environmental Pollution 122*, 105-117.

[10] Lin, M.D., Lin, Y.Ch. (2002). The application of GIS to air quality analysis in Taichung City, Taiwan, ROC. *Environmental Modeling & Software 17*, 11-19.

[11] Bachman, W., Sarasua, W., Hallmark, S. and Guensler, R. (2000). Modeling regional mobile source emissions in a geographic information system framework. *Transportation Research Part C 8*, 205-229.

[12] Wang, L., Parker, D.B., Parnell, C.B., Lacey, R.E., Shaw, B.W. (2006). Comparison of CALPUFF and ISTST3 models for predicting downwind odor and source emission rates. *Atmospheric Environment 40*, 4663-4669.

[13] Mower, B. (1998). A multiple source approach to acute human health risk assessment. *Waste Management 18*, 377-384.

[14] Mitchell, A. (2005). *The ESRI Guide to GIS Analysis. Volume 2: Spatial Measurements & Statistics*. Redlands, California: ESRI Press.

[15] MacDonald, A. (2001). *Building a Geodatabase*. Redlands, California: ESRI Press.

In: Environmental Modeling with GIS
Editor: Lubos Matejicek, pp. 71-83

ISBN: 978-1-60876-363-4

Chapter 5

SPATIAL MODELING AND OPTIMIZATION OF MUNICIPAL SOLID WASTE COLLECTION IN URBAN REGIONS

Lubos Matejicek

Institute for Environmental Studies Charles University in Prague, Faculty of Natural Science, Benatska 2, 128 01, Prague, Czech Republic

ABSTRACT

Growing production of municipal solid waste in the recent years caused by increasing global levels of economic activity needs more sophisticated management tools for its collection. Despite some local efforts to reduce the waste, disposal in landfills is still the most predominate way. Considering to restrictive environmental regulations, more efficient and sustainable systems of municipal solid waste management are carried out to protect our living environment. They deal with the waste sorting, which is mainly focused on categories represented by paper, mixed plastics, mixed glass, mixed waste, hazardous waste and metals. Thus, specific transport vehicles and optimized routes are needed to manage the collection of all the waste categories from local sites to landfills, incineration plants, collecting yards and recovery sites. This contribution presents the waste collection methods based on spatial analysis and operations research. The spatial analysis is focused on development of the road network datasets for waste collection, finding the best routes, defining the service area and creating the origin-destination cost matrix in geographic information systems (GISs). The identified routes and local service areas together with the cost per unit and the total amount of the transported waste form data inputs into operations research. The transportation problem in linear programming is used to optimize the total cost for transportation of each waste category in the framework of municipal solid waste collection. As an example, the urban areas of Prague are included into a case study to illustrate spatial modeling in the GIS environment and optimization focused on collection networks. The individual collection sites are grouped into the urban areas. Waste production in each urban area is estimated in dependence on available data from reports about the living environment in Prague. The locations of target collection points represented by recycling places, incineration plants and landfills are based on larger urban districts. The developed GIS project can manage various scenarios, which indicate different total costs. In spite of its actual lower total cost, the landfill will become more difficult to implement due to the increasing cost caused by limited land availability and environmental community opposition. On the other hand, the recovery/recycling will become more important because of its minimal negative interaction with the living environment. In order to minimize the costs and degradation of our living environment, spatial modeling extended by methods focused on linear programming can assist to find suitable waste management scenarios for collection of municipal solid waste in urban areas.

Keywords: *Municipal solid waste collection; operations research; GIS; spatial modeling.*

1. INTRODUCTION

In the last hundred years, the rapid urban growth has exerted heavy pressures on resources located in the neighborhood, and resulted in serious environmental problems. A few last decades of the twentieth century saw a dramatic increase in the production of urban solid waste caused by unprecedented global levels of economic activity. Thus, urban population has been consuming land and producing wastes at an increasing rate.

Nowadays, solid waste management represents a problem of major relevance for all societies. The growing focus on environmental issues and the goal of sustainable development have moved the problem concerned on solid waste to the forefront of the public agenda. Such a consciousness has led to the development of pollution control techniques that make use of new remediation technology. Consecutively, legislation and regulations have been introduced in the local scales and in the national level to guide waste management. In addition to described management rules and techniques, strategies for sustainable waste management have emphasized the need to minimize waste production, reduce the landfill methods and increase waste recovery. Despite general agreements on the best practices for sustainable development of urban solid wastes, the goal of sustainable waste management seems far from being reached. Generally, the net waste production increases as population grows and the per capita production of waste is also increasing. Disposal in landfills is still predominant destination and the amounts of wastes being recovered are generally small. The situation is different in each particular country considering on the local costs, community opposition and restrictive environmental regulations. In general, landfill is the waste destination method that is characterized by the largest demand for land, which is a decreasing resource in urban areas [1].

Finding acceptable strategies to cope with solid waste management is becoming the inevitable task, owing to the increasing awareness of environmental issues by urban community. In the framework of European region, legislation is carried out to recommend the development of the local management plans, which accent prevention, waste reduction and recovery. In the Czech Republic, the new legislation for municipalities, the permanent increase of solid waste production and landfill shortage have encouraged the development of incineration technology and recycling programs. The methods are accommodated to take into account current economic and environmental conditions.

The aim of this work is to present the application of spatial modeling and optimization focused on municipal solid waste collection in urban regions. Spatial modeling of municipal solid waste collection represents a part of the research related with solid waste management. Principles dealing with a collection network are considered in the context of geographic information systems (GISs). GIS is able to manage spatial data containing collection networks and estimates of municipal waste production. Nowadays, optimization of collection networks can be solved with existing GIS capabilities in an efficient way, but extensions are necessary to achieve operations research functions. Generally, GIS by itself describes the spatial environment of the objects that are related to the municipal solid waste collection. Operations research optimizes the functioning of processes that are concerned on waste management. Thus, GIS and operations research need to operate together. In this case, GIS can serve as a common data framework and an analysis base for operations research. In spite of that the former GISs and operations research software evolved separately, which caused different data structures, the current GISs support a wide range of data formats, contain various case oriented extensions, and are supported by a number of programming development tools. There are a few concepts of integration both the systems on the computer platform [2, 3]. The low level integration is based on exchanging files. The GIS is mostly used to prepare the data inputs, which are then passed to another software package, and after processing, the results are returned backward to the GIS tools for display and spatial analysis together with other map layers. The higher level of integration includes sharing programming standards such as Microsoft .NET or Java that enable common running of functions through single scripts. The final step of integration can represent object oriented programming applications based on software development tools and programming libraries that cover both the GIS functionality and a selected range of the operations research methods. The higher level of integration, the more efficient running and the more experience needed to integrate all the software tools. For the aim of the presented study, the scope of the GIS integration is limited to the selected aspects of operations research focused on transportation problems.

2. Waste Management Models

Construction of any municipal solid waste collection management system in the GIS environment requires the development of a spatial data model that contains data structures of vector objects, raster layers or digital terrain models extended by attributes and topological rules. Consecutively, the data structures can be modified to support optimization methods, which are focused on particular tasks, but the nature and solution of their mathematical equations are the most critical features, rather than spatial modeling. Thus, performing these tasks within GIS requires the expression in GIS language and programming extensions.

Originally, the distinction between the spatial models in GIS and the tools for solution of operations research methods resulted in the separate development of optimization tasks in waste management information systems. Thus, waste management information systems are developed as standalone software tools and implemented models are usually based on methods for analysis of on-site collection systems of waste materials separated at the source for recovery [4, 5, 6]. The analyses can identify potential separation strategies and calculate amounts of materials collected for recovery. Finally, modeling systems offer calculation of effects of the individual separation strategies on costs and emissions. A number of time series analyses and forecasting scenarios are based on statistical methods [7]. For example, the seasonal AutoRegressive and Moving Average (sARIMA) methodology, dealing with short and medium term forecasting, can make predictions based on time series data of solid waste generated quantities.

Some case studies use methodology combining standard systems engineering with operations research methods [8, 9]. The optimization is mainly focused on the minimization of waste transportation costs in major urban areas and the comparison of waste collection strategies. The optimal number of landfills and treatment plants is used to be planned together with the optimal quantities and the characteristics of the refuse that are sent to treatment plants, to landfills and to recycling. A wealth of information on municipal solid waste management is included in the US EPA reports accessible via official web pages. In addition to the US EPA reports and guidelines, huge sources of information are published in the case studies that assist for municipal solid waste collection and optimization in specific regions.

A number of specific software tools have been developed in the last decade in co-operation with the state authorities and the local agencies. The computer programs focused on municipal solid waste collection are often developed as database applications that can manage information about amounts of municipal solid waste and its categories together with the regional demographic attributes and the economic factors. Created reports extended by basic exploratory statistical outputs are used for decision-making processes on the local and the national level. More and more actual data and software tools are accessible via the Internet.

3. Modeling and Optimization within GIS

Using GIS in the framework of waste management systems is documented by an increasing number of case studies in the latest years. Originally, GIS was used to locate potential waste disposal sites, to estimate amounts of waste in urban areas, and to find the best routes for waste collection [10]. Moreover these tasks, representation of spatial objects within a spatial domain by the data linked to the layers of points, lines, polygons and grid cells can be involved into more complex projects that enable to produce new layers based on existing data, analytical functions, numerical modeling and other external data inputs.

Recent GISs include a number of case oriented modules and extensions that are focused on partial tasks. Thus, in addition to basic modules dedicated to manage spatial data, the case oriented extensions are included to process advance tasks in spatial analysis, 3D modeling, geostatistical analysis, network optimization and spatial visualization [11, 12, 13]. For example, ArcGIS Network Analyst can assist to provide network-based spatial analyses that include routing, travel directions, closest facility, and service area analysis. The range of network analyses depends on the accessible spatial dataset that contains a range of attributes holding traffic restrictions and road conditions in dependence on specific times of the day. Thus, a number of spatio-temporal analyses can be conducted including: point-to-point routing, route directions, optimum route, service area definition and closest facility. If the functionality of GIS and its extensions are not sufficient, the GIS development tools assist to build and deploy custom applications or to extend the GIS architecture and spatial data models.

4. Operations Research

The terms operations research, management science or industrial engineering are closely related and often used synonymously. It is caused by using the interdisciplinary branch, which uses methods related to mathematical modeling, statistics, optimization, stochastic processes, graph theory, decision analysis and simulation [14]. Thus, because of the computational nature of these fields, operations research is also related to computer science. The aim of methods is usually to arrive at optimal or sub-optimal decisions in complex problems, which are concerned with searching the maxima or minima of the defined criterion. The searching is based on numerical algorithms that are represented by a large number of repeating steps, which lead to the time consuming computer processing. Thus, good or sub-optimal solutions are often sufficient [15].

The methods used to minimize the costs of municipal solid waste collection are in the rank of linear programming problems that involve the optimization of a linear objective function, subject to linear equality and inequality constrains. The most intuitive form of describing a linear programming problem consists of a linear function to be maximized:

$$c_1 x_1 + c_2 x_2 + \ldots + c_n x_n \cong \max,$$

where the constants c_1, c_2, ..., c_n are mostly the unit shipment cost and the non-negative variables x_1, x_2, ..., x_n are the quantities of the commodity that is shipped.

The other part is represented by problem constraints in the following form:

$$\begin{aligned} a_{11}x_1 + a_{12}x_2 + \ldots + a_{1n}x_n &\le b_1 \\ a_{21}x_1 + a_{22}x_2 + \ldots + a_{2n}x_n &\le b_2 \\ \ldots \\ a_m x_1 + a_m x_2 + \ldots + a_{mn}x_n &\le b_m \end{aligned},$$

where the constant a_{ij} is the unit capacity of the ith source for the production of the jth commodity and the constants b_1, b_2, ..., b_m are the whole capacities for production of individual commodities. Every linear programming problem can be converted into a dual problem, which consists of a linear function to be minimized and the non-equalities in the problem constraints to be reversed. The method of linear programming implemented in the framework of municipal solid waste collection is the so-called transportation problem. It involves an assignment that considers several origins and destinations with the goal to optimize a solution at the minimal cost by minimizing transport costs with fixed origins and destinations. It considers linear transport costs based on estimates of known origins and destinations. The transportation problem can be expressed in the form of the table:

x_{11} (c_{11})	x_{12} (c_{12})	...	x_{1n} (c_{1n})	a_1
x_{21} (c_{21})	x_{22} (c_{22})	...	x_{2n} (c_{2n})	a_2
...	...		...	...
x_{m1} (c_{m1})	x_{m2} (c_{m2})	...	x_{mn} (c_{mn})	a_m
b_1	b_2	...	b_n	

In this case, there are m origins represented by resources of the municipal solid waste, and n destinations represented by landfills, incineration plants, collecting yards and recovery sites in the city districts. The quantities a_1, a_2, ..., a_m of each municipal waste category are transported from each of the m origins and received in the amounts b_1, b_2, ..., b_n at each of the n destinations. The constant c_{ij} is the transportation cost (in currency) per unit transported (in cubic meters or tons) from the ith resource to the jth destination place. Likewise, the variable x_{ij} represents the transportation amount (in cubic meters or tons) from the ith resource to the jth destination place. The aim of the optimization in the framework of the transportation problem is to minimize the total transport costs between a set of origins and destinations. It is assumed that the total amount of units produced and demanded is the same, which implies the equilibrium. But, any supplementary unit produced has not to be transported if there is no supplementary demand and any supplementary demand has not to be transported if there is no supplementary supply.

The transportation problem can be expressed by the mathematical model:

$$\text{Total cost} = \sum_{i=1}^{m}\sum_{j=1}^{n} c_{ij}x_{ij} \cong \min$$

$$\sum_{j=1}^{n} x_{ij} = a_i, \quad i = 1,2,...,m \qquad ,$$

$$\sum_{i=1}^{m} x_{ij} = b_j, \quad j = 1,2,...,n$$

$$x_{ij} \geq 0, \quad i = 1,2,...,m;\ j = 1,2,...,n$$

where the first equation represents the linear objective function and the rest the constraints.

Generally, the tasks in linear programming can be solved by a simplex method, but various other algorithms are in use for special cases of linear programming problems and derived methods. Their advantage resides in either a simpler calculation, or in a considerable reduction of temporarily recorded information. For example, the solution by one of the heuristic methods is based on the table introduced earlier. Step by step, the cells having the lowest transport costs are selected and the highest possible number of units in dependence on the capacity of the origin and the demand is assigned.

In each step, the corresponding transport amount must be subtracted from the related origin and the related destination for the respective row and the respective column. The iterations continue until all surpluses are used and all demands are answered. The total cost is calculated by the linear objective function. The solution based on the introduced procedure is often near the optimal solution, but other methods are needed to rich the optimal solution.

Linear programming methods, including the transportation problems, are mostly implemented as standalone software systems. Thus, in case of using GIS, a suitable level of sharing is needed to solve the complex projects containing the GIS functionality and the linear programming methods.

5. Display and Visualization of Operations Research Data in GIS

GISs mainly dedicated to manage spatial data models contain a wide range of methods that support display and visualization of transportation networks [16]. In case of linear programming, especially the transportation problems, basic GIS tools can be used to display the route network together with other thematic map layers. The case oriented extensions process advanced network analysis based on network datasets extended by connectivity rules. In addition to processing, thematic maps are generated and exported into various formats. For example, ArcGIS Network Analyst is able to carry out network analysis of spatial datasets created in the environment of ArcCatalog and ArcMap. After processing of analytical tools, the results are included into spatial data as new layers and are displayed in layout views. Then, another tool, ArcGIS ArcScene, can be used to create the visualization outputs.

6. A Case Study

As an example, the optimization of municipal solid waste collection in the urban regions of the capital Prague (area: 496 km^2; total population in December 2005: 1,181,610) is carried out to demonstrate the implementation of the transportation problem in the GIS environment. Figure 1 illustrates the central part of the city captured from satellite images. Figure 2 shows the basic map of the area represented by the individual urban areas and the street network. In case of the capital Prague, the data on municipal waste are aggregated and processed by an information system operated by the Centre for Waste Management. The overall reports and case studies are used by the Ministry of the Environment of the Czech Republic and by the Prague City Hall. Data collection, data processing and data reporting are provided in accordance with the new legislation for municipalities. The municipal waste is sorted into a few categories: paper and cardboard, colored glass, clear glass, mixed plastics, beverage boxes, bulky waste, hazardous waste, ferrous and non-ferrous metals, demolition waste, electro-technical waste, waste from care for greenery, wood waste and tires. The waste disposal is based on the kerbside collecting system or the delivery system for specific municipal waste categories. The schema of the major municipal solid waste collection subsystems is shown in Figure 3.

The sorted waste collection focused on paper and cardboard, glass and plastics is mostly provided by means of the kerbside collecting system, in which the sorted waste is turned into collecting vessels. The collection is managed by the city hall in cooperation with the operating companies. Mixed waste forms the influential part of the municipal waste in the central part of the city and on the territory of blocks of flats. The extra charges from citizens are needed to operate the disposal of the municipal mixed waste.

The mixed bulky waste is collected into large-capacity containers. The containers are allocated to the city places depending on their respective population to prevent uncontrolled dumpsite formation. The placement of containers is managed by the city districts. The bulky waste can be also turn into the collecting yards, which are operated by the city hall. The collecting yards enable to return municipal waste in a larger quantity and in a wide range of other categories. The extra waste management system is represented by the hazardous waste collection that is mostly focused on collection of toxic materials, medicines, batteries, accumulators and electronic subsystems. The waste collection is provided by a few ways: stationary and mobile collection points, pharmacies (drugs, medicines and mercury filled thermometers), schools and city offices (spent batteries and accumulators). Disposal of each municipal waste category is provided separately by the special collection equipment, transport vehicles and final treatment.

Figure 1. The capital Prague: the satellite image of the center and the surrounded areas (captured from Google Earth in September, 2007).

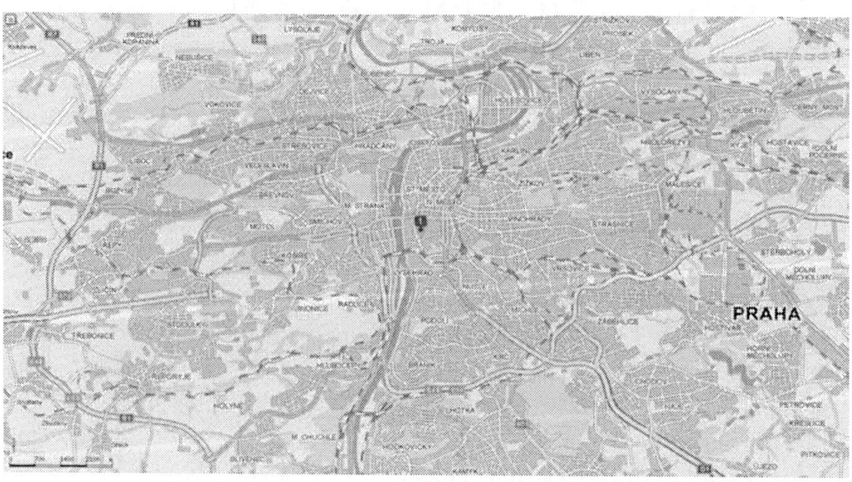

Figure 2. The capital Prague: the basic map of the center and the surrounded areas (captured from the website Mapy.cz in September, 2007).

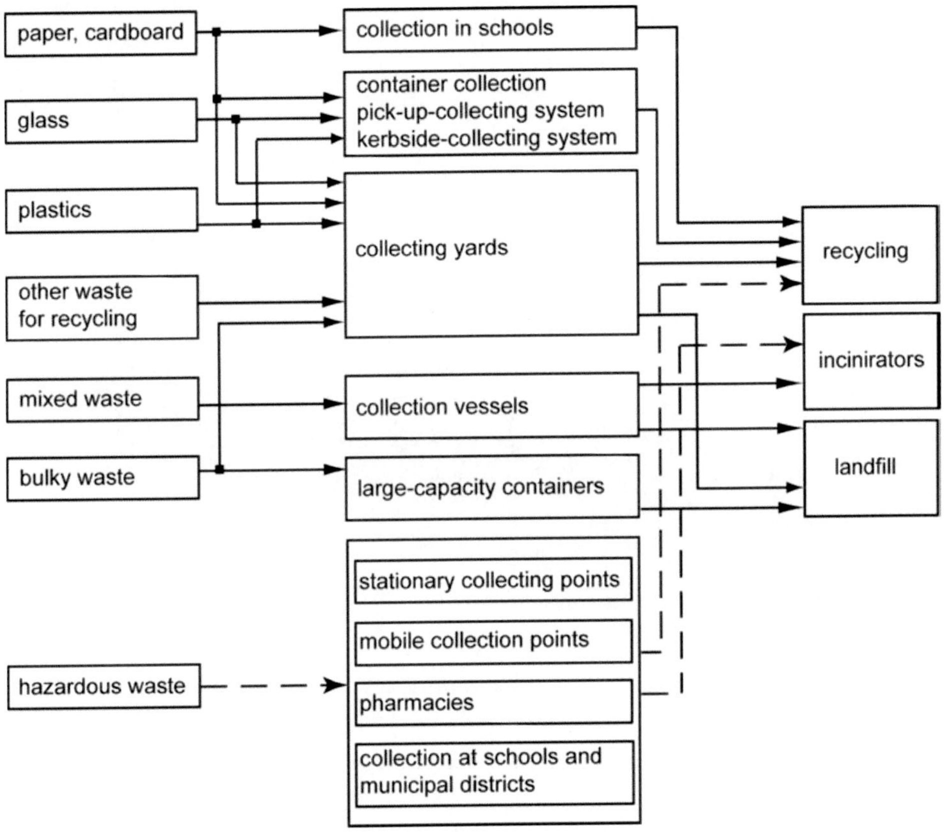

Figure 3. A basic schema of the municipal solid waste collection (based on information from the yearbooks of Prague [17] and local technical reports mostly accessible via the Internet).

6.1 Building of the Spatial Data Set in the Environment of GIS

In case of using GIS for network analysis and visualization, spatial data have to be converted into the spatial data set (ESRI's geodatabase or a specified set of files). The spatial dataset contains geographic locations of spatial objects and their topological rules, which are necessary for network analysis. To demonstrate the optimization of municipal solid waste collection in a simple way, the collection places for each category are reduced into the central points representing individual urban areas as the resources. The resources are divided into a few basic categories: paper, mixed plastics, mixed glass and mixed waste. Other categories like hazardous waste, bulky waste and metals are not included into the set, because their collection management is different. The set of target points is represented by recycling places, incineration plants and landfills. In order to solve the transportation problem and present the principles of waste collection in the GIS environment, the number of target points is reduced to simplify the structure. As an example, the GIS project is developed to demonstrate spatial data management, network analysis and sharing operations research data. The waste resources and waste targets are included into the GIS map layers and illustrated together with the city road network in Figure 4.

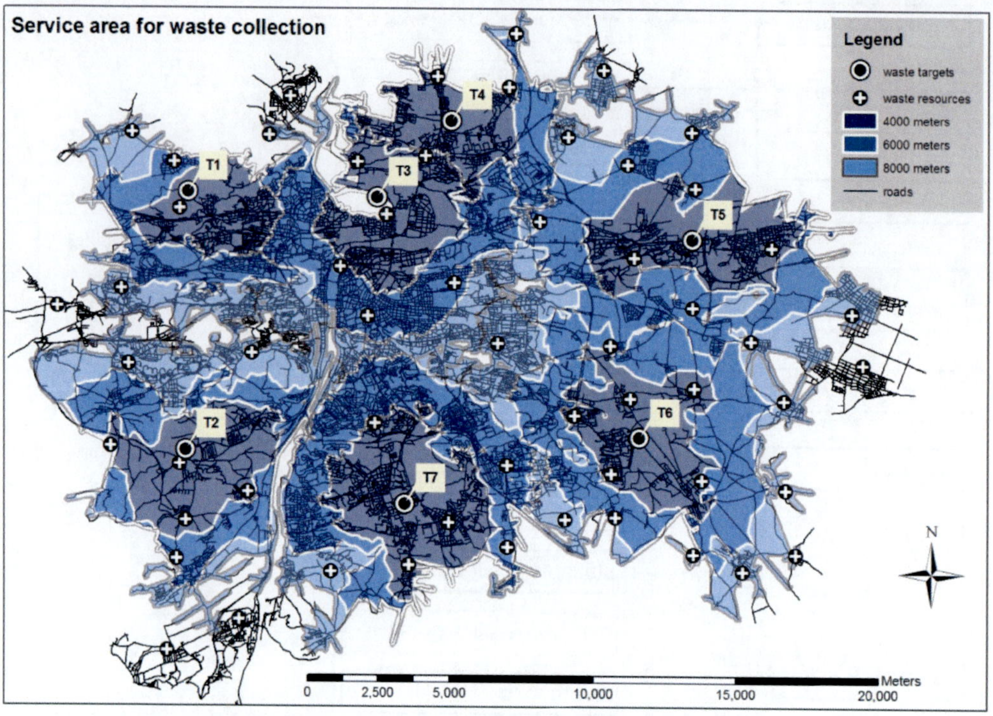

Figure 4. The schema of waste resources and waste targets extended by service areas.

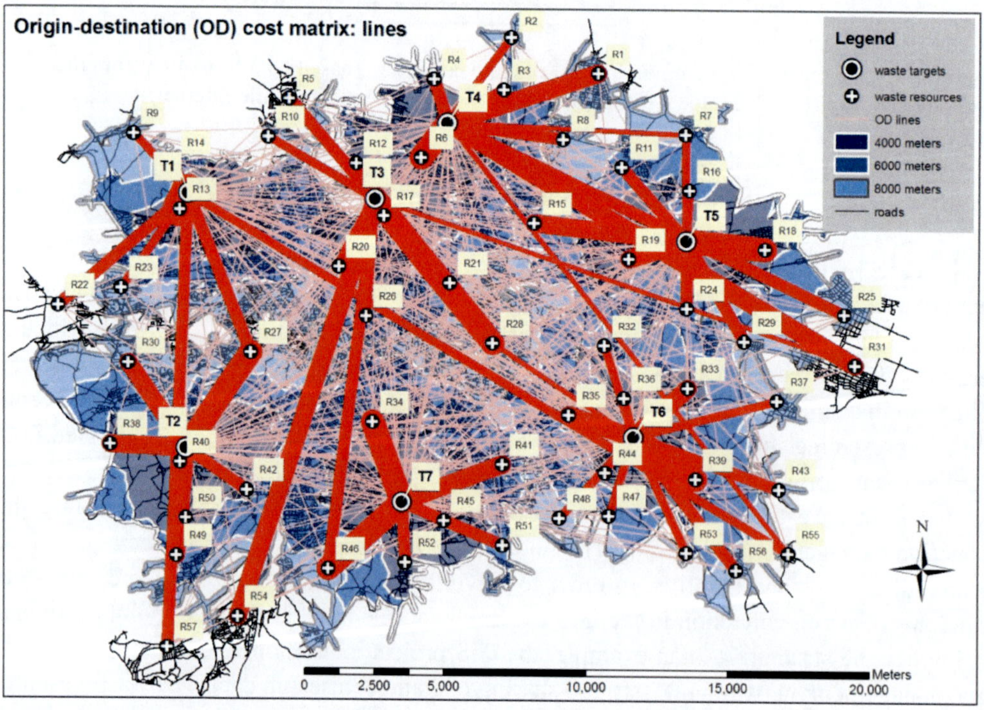

Figure 5. The schema of waste resources and waste targets extended by OD lines.

The GIS extensions can solve a number of network analyses based on finding the best route among a set of resources or finding the closest target points. In case of this example, spatial analysis is used for calculating the service area and creating the Origin-Destination (OD) cost matrix. The service areas around the target points for selected road distances (4000, 6000, 8000 meters) are displayed in Figure 4. The GIS network extensions (NetworkAnalyst in ArcGIS) is used to create a series of polygons representing the distance that can be reached from target points within a specified road distance. The OD cost matrix is represented by a series of the connection lines in Figure 5. The matrix is used as the input for the transportation problem. Each connection line includes a number of attributes. The road distance between the resource point and the target point is used to estimate the transportation cost per unit transported from the ith resource to the jth target. The estimated transportation costs together with the estimated quantities of each municipal waste category at the resources are transferred into the numerical module for solution of the transportation problem. After the calculation, the optimized transportation amounts between the resources and targets are backward transferred into the GIS project. The solution based on optimization of the total transport costs between the set of resources and the set of targets is illustrated in Figure 5.

The width of the OD lines represents the relative transport amounts among the individual pairs. The thin connection lines on the background show the whole set of pairs, which, in this case, do not serve for transportation.

6.2 Sharing Data between Operations Research Tools and GIS

The presented case study demonstrates calculation of the transport problem by sharing data between GIS and the specialized programming module. In case of the selected city, the datasets contain 57 resources and 7 targets for each municipal waste category. The estimates of attributes in the table (c_{ij}; R_i; T_j) are based on GIS spatial data (road distance, size of each area). The transportation problem is expressed by the linear objective function:

$$\text{Total cost} = \sum_{i=1}^{57}\sum_{j=1}^{7} c_{ij}x_{ij} \cong \min$$

and by the constraints:

$$\sum_{j=1}^{57} x_{ij} = a_i, \quad i = 1, 2, ..., 57$$

$$\sum_{i=1}^{7} x_{ij} = b_j, \quad j = 1, 2, ..., 7$$

$$x_{ij} \geq 0, \quad i = 1, 2, ..., 57; j = 1, 2, ..., 7$$

The output variables (x_{ij}) represent the transportation amounts. The solution based on heuristic methods is expressed in the table:

Waste category: (paper, mixed plastics, mixed glass and mixed waste)							
$x_{1,1}$ ($c_{1,1}$)	$x_{1,2}$ ($c_{1,2}$)	$x_{1,3}$ ($c_{1,3}$)	$x_{1,4}$ ($c_{1,4}$)	$x_{1,5}$ ($c_{1,5}$)	$x_{1,6}$ ($c_{1,6}$)	$x_{1,7}$ ($c_{1,7}$)	R_1
$x_{2,1}$ ($c_{2,1}$)	$x_{2,2}$ ($c_{2,2}$)	$x_{2,3}$ ($c_{2,3}$)	$x_{2,4}$ ($c_{2,4}$)	$x_{2,5}$ ($c_{2,5}$)	$x_{2,6}$ ($c_{2,6}$)	$x_{2,7}$ ($c_{2,7}$)	R_2
...	...	...	...	...	...	...	...
$x_{57,1}$ ($c_{57,1}$)	$x_{57,2}$ ($c_{57,2}$)	$x_{57,3}$ ($c_{57,3}$)	$x_{57,4}$ ($c_{57,4}$)	$x_{57,5}$ ($c_{57,5}$)	$x_{57,6}$ ($c_{57,6}$)	$x_{57,7}$ ($c_{57,7}$)	R_{57}
T_1	T_2	T_3	T_4	T_5	T_6	T_7	

CONCLUSION

The synthesis of GIS and operations research, likewise dynamic modeling [18], in the framework of waste management focused on municipal solid waste collection has the potential to create a new base for optimization tools that assist to solve waste treatment and waste disposal in a more efficient way. Utilizing GIS and development operations research tools on it is making cost effective and highly efficient systems, which can be implemented into urban planning and decision-making processes. Using GISs for brings possibilities of using standard spatial data formats, implemented analytical methods and visualization tools. The outputs integrated into the map layers and managed by GIS servers offer new ways of sharing information on the local level of the city regions. The operation research methods have been using for a few decades. Despite time consuming numerical algorithms for solution of the optimization tasks, the progress in computer technology and new GIS software can be extended by the case oriented modules focused on operations research methods including linear programming.

The attached case study illustrates the potential possibilities for optimization of waste management tasks. In addition to the standard solution of the transportation problem, the setting of the higher transportation costs between the resources and the targets represented by disposal in landfills or incineration plants enable to carry out the new scenarios, which are more sustainable and protect our living environment. Thus, the extra costs can be used for recycling and building of the new recovery sites. As an example, only one category of municipal solid waste is selected to display the service areas and the OD lines in the previous figures, because the vehicles use the same road network for other waste categories. However, the transportation amounts between the set of resources and the set of targets depend on each category.

The solution of optimization tasks for each category of the municipal solid waste offers sharing of the data outputs together with other map layers in the GIS environment, which help in urban planning and environmental protection. New scenarios can be created by modification of data inputs into transportations problem in order to find the suitable solution.

ACKNOWLEDGMENTS

The spatio-temporal data and operations research methods have been processed by ArcGIS and programming software tools in the GIS Laboratory, Faculty of Natural Science, Charles University in Prague.

REFERENCES

[1] Leao, S., Bishop, I. and Evans, D. (2001). Assessing the demand of solid waste disposal in urban region by urban dynamics modelling in a GIS environment. *Resources, Conservation and Recycling 33*, 289-313.

[2] Maidment, D.R. (1996). Environmental Modeling within GIS. In M.F. Goodchild, L.T. Steyaert, B.O. Parks (Ed.), *GIS and Environmental Modeling: Progress and Research Issues* (pp. 315-323). Fort Collins: GIS World, Inc.

[3] Fedra, K. (1996). Distributed Models and Embedded GIS: Integration Strategies and Case Studies. In M.F. Goodchild, L.T. Steyaert, B.O. Parks (Ed.), *GIS and Environmental Modeling: Progress and Research Issues* (pp. 413-417). Fort Collins: GIS World, Inc.

[4] Tanskanen, J-H. (2000). Strategic planning of municipal solid waste management. *Resources, Conservation and Recycling 30*, 111-133.

[5] Choe, C. (1999). An Economic Analysis of Household Waste Management. *Journal of Environmental Economics and Management 38*, 234-246.

[6] Barlishen, K.D. and Baetz, B.W. (1996). Development of a Decision Support System for Municipal Solid Waste Management Systems Planning. *Waste Managementand Research 14*, 71-86.

[7] Navarro-Esbrí, J., Diamadopoulos, E. and Ginestar, D. (2002). Time series analysis and forecasting techniques for municipal solid waste management. *Resources, Conservation and Recycling 30*, 201-214.

[8] Kulcar, T. (1996). Optimizing solid waste collection in Brussels. *European Journal of Operational Research 90*, 71-77.

[9] Fiorucci, P., Minciardi, R., Robba, M., Sacile, R. (2003). Solid waste management in urban areas Development and application of a decision support system. *Resources, Conservation and Recycling 37*, 301-328.

[10] Muttiah, R.S., Engel, B.A., Jones, D.D. (1996). Waste Disposal Site Selection Using GIS-based Simulated Annealing. *Computers and Geosciences 22*, 1013-1017.

[11] MacDonald, A. (2001). *Building a Geodatabase*. Redlands, California: ESRI Press.

[12] Maguire, D.J., Batty, M., and Goodchild, M.F. (2005). *GIS, Spatial Analysis and Modeling*. Redlands, California: ESRI Press.

[13] Mitchell, A. (2005). *The ESRI Guide to GIS Analysis. Volume 2: Spatial Measurements and Statistics*. Redlands, California: ESRI Press.

[14] Dantzig, G.B. (1963). *Linear Programming and Extensions*. Princeton, Princeton University Press.

[15] Fletcher, R. (1987). *Practical Methods of Optimization*. New York, Wiley.

[16] Maguire, D.J., Batty, M., and Goodchild, M.F. (2005). *GIS, Spatial Analysis and Modeling*. Redlands, California: ESRI Press.

[17] Solc, J. and Pospisilova, J. (2007). Prague-Environment 2006. Prague, J.I.M. Liberec.

[18] Matejicek, L. (2006). Modeling of Water Pollution in River Basin with GIS. In A.R. Burk (Ed.), *Focus on Ecology Research* (pp. 249-270). New York, Nova Science Publishers, Inc.

In: Environmental Modeling with GIS
Editor: Lubos Matejicek, pp. 85-98
ISBN: 978-1-60876-363-4

Chapter 6

SPATIAL MODELING OF AIR POLLUTION BASED ON TRAFFIC EMISSIONS IN URBAN AREAS

Lubos Matejicek
Institute for Environmental Studies, Charles University in Prague
Benatska 2, 128 01, Prague, Czech Republic
Zbynek Janour
Institute of Thermomechanics, Academy of Sciences
Dolejskova 5, 182 00, Prague, Czech Republic
Michal Strizik
Laboratory of Risk Research and Management, Technical University of Ostrava
Lumirova 13, 700 30, Ostrava, Czech Republic

ABSTRACT

Road traffic becomes a dominant source of air pollution in urban areas. The emissions of inorganic compounds and volatile organic compounds caused by motor vehicles have increased trends, particularly in urban areas, which together with other emitted compounds can cause human health problems in a long-time period. Procedures for estimating of road traffic emissions in the USA and in the European region are based on guidelines and recommendations. In order to implement the guidelines, decision-making tools are necessary to provide information in an easy understandable form. Thus, the spatial information system is needed to manage spatio-temporal data, provide analysis, solve numerical models and visualize the results. Generally, spatial data include street networks and monitoring networks. Temporal data are represented by changes of traffic intensity and by time series of measured pollutants. Geographic information systems (GISs) extended by environmental modeling tools offer advance estimates of traffic emissions improved by the digital terrain data and by the wind flow effects. The estimates of air pollution in the surrounded areas can be based on spatial interpolation. In a wide range of spatial interpolations, the deterministic techniques or the geostatistical methods are used in dependence on the input data and the results provided by the exploratory spatial data analysis (ESDA). Each spatial interpolation attached to the map layer defines concentration levels at a specified time period. The time series of map layers can demonstrate the variability of the air pollution distribution, which brings a new insight into this research.

In addition to mapping of the urban air pollution in dependence on the traffic intensity and the measured concentrations, numerical modeling and simulation in wind tunnels are presented as important tools for exploration of the dispersion and the transport of pollutants caused by wind flows and other effects. A number of software tools based on Gaussian dispersion principles are mentioned in the framework of US EPA guidelines.

As a case study, spatial modeling of air pollution is focused on the urban area of the city of Prague, because a number of motor vehicles registered on the Prague territory is growing. In addition to the registered vehicles, the specific phenomenon in the central Europe is represented by the abrupt increase in traffic of trucks. In spite of new methods for reduction of emissions from motor vehicles, urban development and successive reconstructions of existing roads cause other emissions. Thus, the street network also becomes a set of line sources of stirred up suspended particulates generated by the passing vehicles. These dust emissions, so-called secondary dust, are estimated and partially validated by measurements of the fraction PM_{10}. As an example, estimates of nitrogen oxides are carried out by spatial interpolation based on data from the surface monitoring network and the DIAL-LIDAR measurements.

Integration of spatio-temporal data together with environmental modeling tools brings new possibilities to compare spatial interpolations created for individual compounds at the high temporal and spatial resolution. In the framework of the GIS, mapping of air pollution and human exposures in streets, at workplace locations and residence addresses can help to reduce traffic emissions by optimization of transport scenarios.

Keywords: *Road traffic; air pollution; spatial modeling; GIS.*

1. INTRODUCTION

Transport-related emissions are rising and there is increasing consensus that the growth in motorized vehicles in urban areas is incompatible with health, and greenhouse-gas emissions. Traffic intensity and air pollution adverse health effects and overall environmental degradation. Mostly, transport is the only sector for which current emissions are higher than in the last seasons. In the USA, passenger-car emissions and emissions from trucks are increasing. In Europe, freight transport has more than doubled since 1970. In eastern parts of Europe, the 1990s are characterized by a major decline in rail transport and a concurrent increase in car and truck use. The escalation of urban transport is also significant in Chinese cities in the last decade. Thus, the transport, especially in urban areas, is widely recognized as a significant and increasing source of air pollution.

Many reviews are focused on regional and global impacts of transport emissions of air pollution, but mostly deal with the impact of emissions on local urban air quality, and consider only emissions related to the road transport. In addition to these reviews, some reports based on data from protection agencies (US EPA or European Environmental Agency) cover air and sea transport. There is also an emphasis on assessing possible solutions to environmental problems related to transport. But, too little action is taken on to reduce the environmental impacts. For all that the catalytic convertors are being fitted to every new gasoline-engine car, which is followed by similar developments in diesel emissions, impacts of road transport on urban air quality is still growing.

In spite of that the mode of transport is dependent on the selected area, which can include also ships and aircrafts, the attached case studies are focused on road traffic and its effects on air pollution in urban areas. Road transport is characterized from other sources of air pollution in that the pollutants are released in very close proximity to human receptors. It also reduces the opportunity for the atmosphere to dilute the pollutants. In addition to the close proximity, the roads in the city centre have buildings alongside that shelter the roads and reduce the wind speed at the sources of emissions to that on open roads. But this increase of transport emissions has not significant impact on damage human health because people usually spend larger amounts of time indoors [1]. In case of the buildings that line city streets, the indoor concentrations of pollutants are mainly dependent on outdoor concentrations from road traffic adjacent to windows and doors.

2. Numerical Modeling Focused on Air Pollution

In order to quantify impacts of transport emissions, dispersion of pollutants is studied by using of numerical modeling [2] or physical modeling [3]. The numerical dispersion models deal with simulation of increased concentrations near the major roads, compared to overall background levels of emitted compounds by motor vehicles. Due to the wide range of car classes, many compounds must be taken into account. Mostly, the dispersion modeling is focused on carbon monoxide (CO), nitrogen oxides (NO_x) ozone (O_3), coarse ($PM_{10-2.5}$), fine ($PM_{2.5}$) and ultrafine ($PM_{0.1}$) particles, black carbon (BC), polycyclic aromatic hydrocarbons (PAHs), and benzene. The numerical models are mostly based on Gaussian dispersion principles. Many state agencies and local authorities, especially in the USA, use the modeling program Industrial Source Complex-Short Term, Version 3 (ISCST3) approved and recommended by the US EPA. The ISCST3 implements a set of Gaussian plume-based models that are used to calculate downwind concentrations from point, line, and area sources. In dependence on a spatial scale of dispersion modeling, the roads are represented by line sources, and the crossroads, parking places, bus stops and other transport service places can be represented by point or area sources. While the ISCST3 is developed as the steady-state air emission modeling system, another dispersion model called CALPUFF represents a generalized non- steady-state air emission modeling system. The US EPA approves to use the CALPUFF as a guideline model for regulatory applications involving long range transport, and for near-field applications with non-steady-state effects. In case of local prediction of air pollution from road transport, many studies focused on the middle spatial scale tasks are based on these environmental modeling systems [4, 5, 6, 7]. The ISCST3 and the CALPUFF are complemented by other computer tools for processing of meteorological data, spatial modeling of buildings, and visualization of simulation results.

While the ISCST3 or the CALPUFF represent modeling guidelance, many dispersion models are directly focused on dispersion modeling of traffic emission sources. For example, the CALINE, based on Gaussian diffusion equations, employs a mixing zone concept for pollutant dispersion over the roadway. The prediction of pollutant concentrations is carried out for receptors located within 500 meters of the roadway. The inputs contain data about emissions sources, meteorology and site geometry. The special options are included for modeling of air quality near the intersections, street canyons and parking facilities. Nowadays, the CALINE together with the ISCST3 and other dispersion models are included in the US EPA model library, guidelines and recommendations.

3. Modeling in Wind Tunnels

In case of physical modeling, the relation between urban three-dimensional morphology and windiness is explored by wind tunnel experiments. Highly idealized city models consisting of blocks of buildings with street canyons are usually used for construction of the physical models. In addition to the physical models based on blocks of buildings, there are many complements for a more realistic simulation in the wind tunnels. Trees with various shapes of crowns are positioned inside the street canyons. Line sources are embedded in the bottom of the streets to release of the tracer gas for simulation of traffic emissions. Traffic-induced turbulence is simulated by rotating belts with thin plates etc. [8]. Physical modeling is used to be complemented by numerical modeling based on Navier-Stokes equations that are extended by tools for simulation of turbulent flows and calculation of boundary conditions. Many case oriented studies oriented on the urban environment are included in the books and papers focused on computational wind engineering [9]. The evaluation of surface flow fields in the urban environment can be extended by vertical flows caused by daytime heating processes in the summer periods and by other phenomena [10]. Thus, the wind tunnel experiments together with numerical simulation can provide the encouraging insight into air flows and offer to predict transport of air pollutants from traffic emission sources.

4. Methods Based on Exploratory Spatial Data Analysis

Due to required huge amounts of data and time consuming numerical calculations, another ways of environmental modeling are often used to predict concentration levels of pollutants in the neighborhood of emission sources. For example, the thematic maps focused on pollutant concentrations over urban areas are mostly based on spatial interpolation that can create the continuous surface of the pollutant concentration, standard errors of predictions and the probabilities that critical values are exceeded. In case of air pollution based on traffic emissions, the inputs into the interpolation tools are represented by estimated levels of concentration at the major emission sources formed by crossroads and road sections or by a surface monitoring network. The spatial interpolation by itself uses sample points at different locations. The sample points include estimates or measurements of selected pollutants.

In order to carry out the proper spatial interpolation, the exploratory spatial data analysis (ESDA) assists to gain a deeper understanding of the pollutant dispersion by creating histograms, Voronoi maps, quantile plots, trend analysis, semivariogram/covariance clouds and crosscovariance clouds. The ESDA is often integrated in the geographic information system (GIS), which allows highlighting the spatial objects on the map layers together with their attributes. Generally, two groups of interpolation techniques are provided by the GIS: deterministic and geostatistical. While the deterministic techniques use mathematical functions for spatial interpolation, geostatistical tools rely on both statistical and mathematical methods and, in addition to creating the surface, can assess the uncertainty of the predictions. The deterministic techniques include Inverse Distance Weighted (IDW) and polynomial interpolations. The geostatistical techniques are based on various kriging models [11]. Implementation of interpolation procedures is usually described in the GIS guidebooks that give more details about setting of the internal parameters [12].

5. Traffic Emission Inventories in Urban Areas

Inventories of atmospheric emissions from transport have been carried out in the framework of research projects and case oriented studies for decades. They are constructed on a range of scales from individual urban areas to countries or continents. The estimates of air pollution are structured in a different way depending on country legislation and on local rules that govern environmental protection. Thus, the dissimilarity between the inventories made in the USA and Europe reflect not only true differences in emissions, but also distinctions related to the way in which the inventory has been produced [13]. In the USA, many guidelines and traffic emission inventories are accessible through the web pages of the US EPA. The American urban areas are larger than the European urban areas that usually include historical centers linked by narrow streets, and the block of flats in the suburbs interconnected by the arterial roads. In the European Union, the emission inventory guidebooks are mostly prepared by the European Environmental Agency. Nearly each year, the updates are published and accessible on the Internet. For example, the ENEP/CORINAIR Emission Inventory Guidebook published in 2007 contains chapters focused on road transport [14]. The guidelines are discussed and applied in a number of studies [15, 16, 17]. Many modeling tools for estimation of traffic emissions in European cities are also based on CORINAIR emission inventories [18, 19, 20].

6. Spatial Estimation of Air Pollution by Laser Scanning

Remote sensing techniques, mostly used for detection and classification of the surface objects [21, 22], are also implemented in the systems for detection of atmospheric pollutants. Information about the pollutants is derived from spectroscopic interaction between radiation and atmospheric matter. The intensity of laser beams is decreased when passing through polluted areas of the atmosphere. In order to obtain the most precise measurements, the DIALs (Differential Absorption LIDARs) are employed in the mobile stations for surface measurements of the pollutants over the city road network. It represents active remote sensing methods where the pulse LIDAR techniques can estimate the distribution of the pollutants hundreds meters over the surface. The laser pulses are emitted in two different wavelengths, resonance and reference. The resonance wavelength is chosen within the absorption band of the measured air pollutant. Thus, the decrease of the intensity is higher than for the intensity of the reference wavelength that is outside of the absorption band. A portion of the laser light of both wavelengths is returned back to the DIAL system by molecules of gases (Rayleigh scattering) and by aerosols (Mie scattering). By comparing signal intensities on both wavelengths, the high resolution detector and processing unit can create the pollutant concentration profile along the trajectory of the laser pulses. The emitor and the detector are managed together by the rotary periscope that emits the laser beams and receives the backscarttered signals from any direction.

Applications of the LIDAR and the LIDAR-DIAL techniques show a growing capacity in environmental monitoring, especially, if spatial and temporal distributions are required [23]. A wide range of atmospheric contaminants, as for example volatile organic compounds, nitrogen oxides, ozone, sulphur dioxide and carbon monoxide, can be detected in the range of a few kilometers [24, 25, 26]. In addition to the measurement of individual pollutants over the urban areas, the horizontal and vertical concentration profiles allow estimating of the mixing height and the relationship between vertical mixing and ground-level pollutant concentrations. Thus, the ground-level concentrations of ozone and nitrogen oxides in urban canyons based on DIALs techniques can be compared with the conventional fixed point sampling of pollution gas concentrations obtained by conventional monitoring networks [27]. The measurements of the vertical and temporal variations based on a few observation periods bring other insight into air pollution transport and chemical interactions during a day period [28]. In order to analyze spatio-temporal measurements of air pollution together with surface spatial objects and traffic sources of pollution, the GISs are often used for the creation of map layers extended by concentration profiles [29]. In addition to the spatial data management and visualization, GISs can remarkably assist in spatial interpolation of the horizontal and vertical profiles derived from DIALs measurements [30].

7. GIS-Based Spatial Analysis and Visualization

GIS environment offers tools for exploratory spatial data analysis that can explore fixed point sampling and DIALs concentration profiles in dependency on surface emission sources, terrain models and other surface objects [31, 32]. Generally, GIS offers to integrate a wide range of spatial data, which can support spatial modeling and in some cases dynamic modeling. Various scenarios of data integration for environmental modeling are described in the book written by Goodchild et al. [33, 34]. In spite of a huge offer of the GIS analytical tools [35, 36, 37], there is still the need to develop new extensions that solve case oriented tasks in the framework of environmental analysis and modeling [38]. In some cases, the file exchange or sharing methods between the GIS and the standalone modeling tools must be used to solve large environmental projects that deal with decision-making processes.

The time series of pollutant concentrations from conventional fixed point sampling networks can be linked to the map layers together with the emission sources. The horizontal and vertical profiles from the DIALs offer to extend standard monitoring techniques by exploring pollutant concentrations over the surface. In case of numerical modeling and wind tunnel studies, the GIS extensions are needed to solve time consuming numerical calculations and to create the interconnection between the GIS and the standalone software tools based on US EPA modeling guidelines and other accessible resources. Nowadays, the RDBMS extended by modules for spatial data management, like ESRI's geodatabase, are used in order to support spatial analysis, numerical modeling and visualization in a more complex way that offers direct integration of spatial data formats, time series and analytical tools [39].

8. A CASE STUDY FOR SPATIO-TEMPORAL ANALYSIS OF AIR POLLUTION FROM URBAN TRAFFIC IN PRAGUE

In order to inform local authorities responsible for controlling road traffic in Prague, monitoring system of air pollution and other risks has been operating for a decade. This monitoring system was developed in the framework of the project, which is shared by the Department of Informatics at the Prague City Hall, by the Czech Hydrometeorological Institute and by the T-Soft company. Aggregated information about air pollution based on the surface monitoring network is presented on the Internet and accessible to public. The ground-level air pollution assessment shares data archived in the database of the Air Quality Information System of the Czech Republic. As an example, Figure 1 illustrates the temporal changes of nitrogen oxides during the time period of a few hours. The upper part contains buttons for interactive data management (functions: refresh, backward, forward, legend and report). The graph can display emission limit concentrations for emergency and regulation that are complemented by a list of selected monitoring stations (accessible on the right side).

Generally, the emission sources generating air pollutants are classified in the framework of the Air Pollution Source Register into a few classes in dependence on risk assessment. In case of the stationary emission sources, there are three classes in the register:

-extremely large and large source combustion processes with the heat-generating capacity above 5 MW and very significant technologies,

-middle-sized source combustion processes with the heat-generating capacity 0.2–5 MW and significant technologies,

-small source combustion processes with the heat-generating capacity under 0.2 MW and less significant technologies.

Monitoring of stationary emission sources is continuously updated and revalued in the framework of implementing of the new monitoring techniques and modeling procedures. The individual class is represented by the mobile emission sources, which are consecutively categorized in dependence on their shape to the point sources (crossroads, portals and tunnel ventilation exhausts, parking places) and the line sources (sections of the urban roads). In order to estimate the emissions from mobile sources in a more precise way, a few projects supported by the Ministry of the Environment of the Czech Republic have been created in the last decade. As an example, the map layer based on a model calculation of traffic emissions is presented in Figure 2. Model calculations take into account the type of vehicles, the composition of traffic flows, the speed, the slope and the estimates of increased emissions generated at a cold start of the vehicle engine. The following pollutants can be included into the tools: PM_{10}, sulphur dioxide, nitrogen oxides, carbon monoxide and hydrocarbons [40].

The spatial interpolations based on data from the surface monitoring stations represent other source of information about living conditions in urban areas. The map layers focused on nitrogen oxides created in the GIS environment with the IDW and with the ordinary kriging are illustrated in Figure 3.

The distribution of the nitrogen dioxide in the street canyon located in the urban area of the city Prague is illustrated in Figure 4. The 3D scenes based on processing of the LIDAR-DIAL data are focused on the arterial street with heavy traffic. The measurements of the concentration located in spatial nodes on the laser lens are interpolated by the IDW and ordinary kriging. As an example, a few time periods are included into the output presentation.

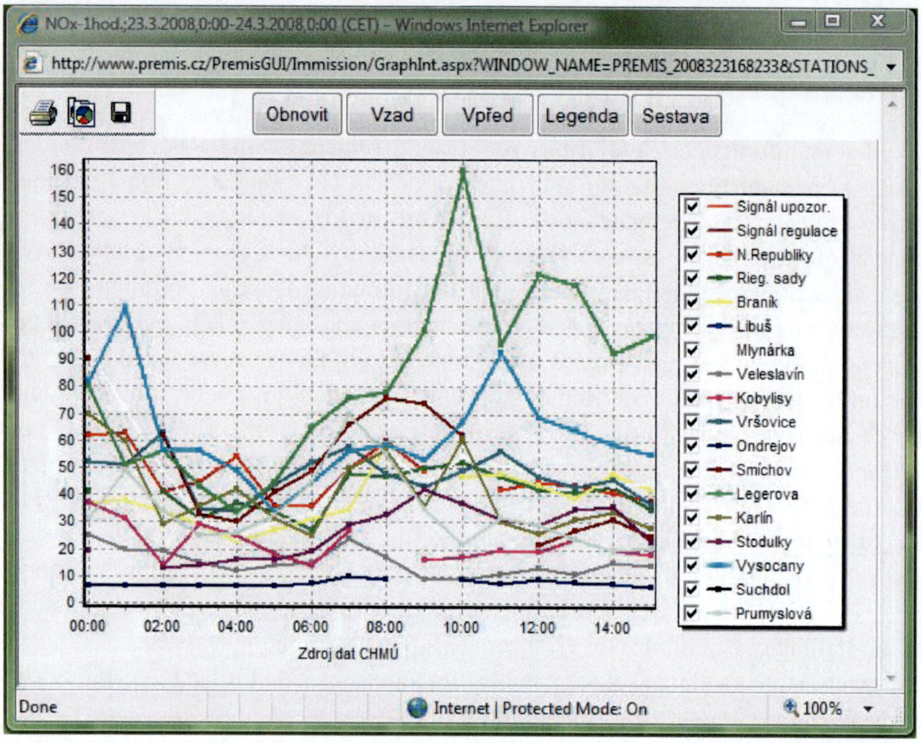

Figure 1. The actual information about air pollution in Prague accessible for public through the Internet (http://www.premis.cz, March-2008).

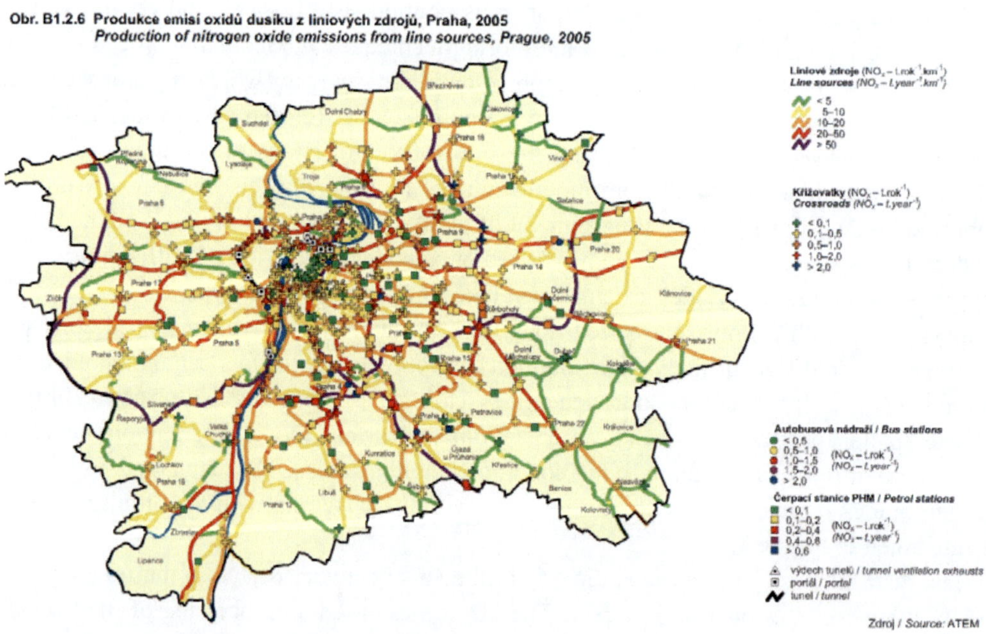

Figure 2. Map layers focused on modeling of nitrogen oxides from line sources (PRAGUE–ENVIRONMENT 2005, MHMP, 2006; http://envis.praha-mesto.cz/, March-2008).

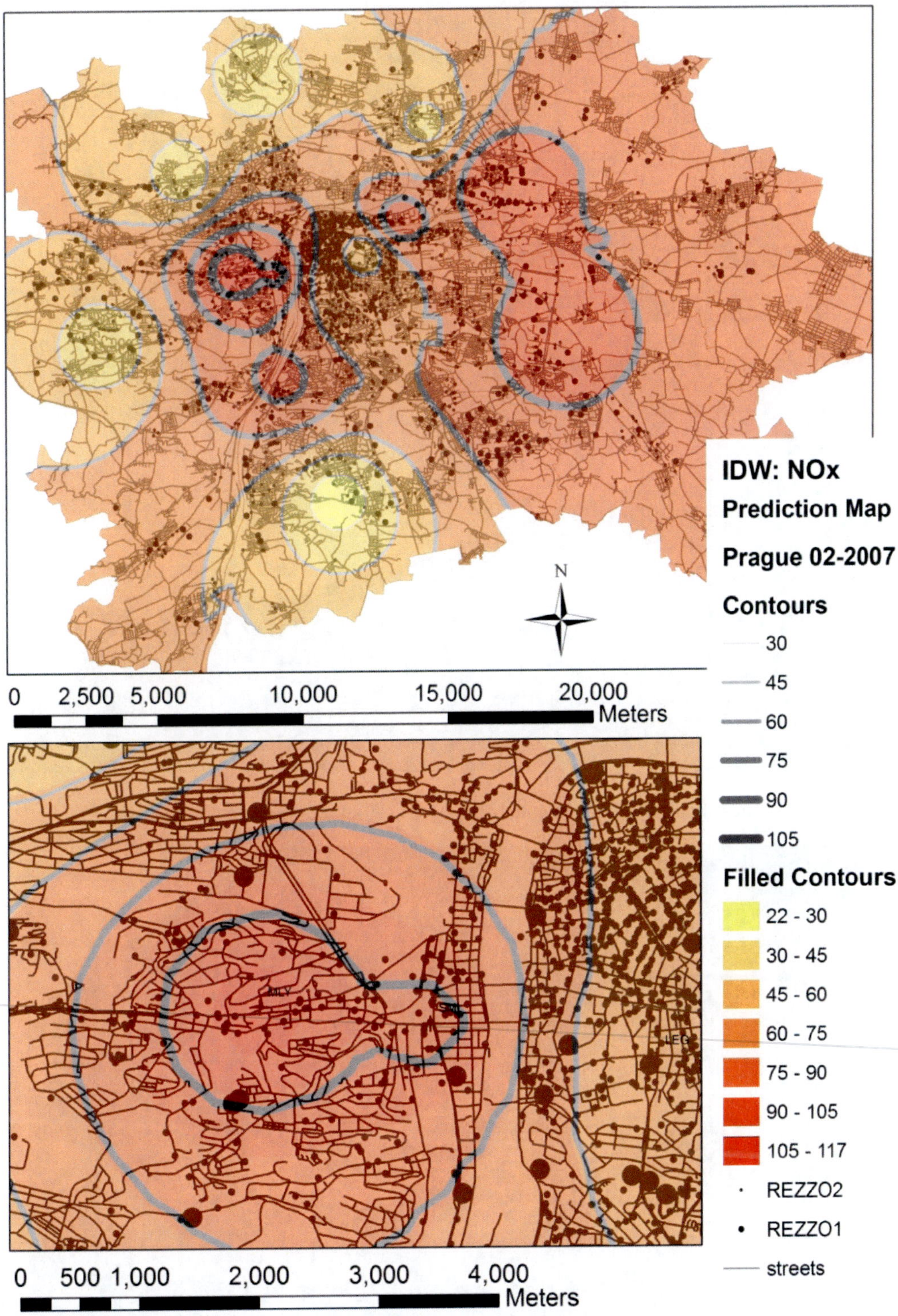

Figure 3. Map layers created in the GIS environment with spatial interpolations (IDW and ordinary kriging) based on data from surface monitoring network, the stationary sources of air emissions are marked in the map layer as points (REZZO1, REZZO2-the Czech guidelines).

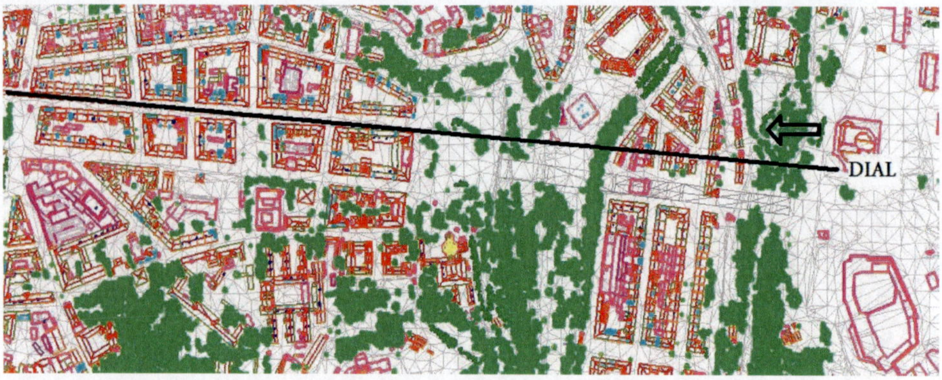

The location of the DIAL's vertical profile in the canyon of an arterial street in Prague.

Visualization of the urban area complemented by the vertical image based on DIAL's data.

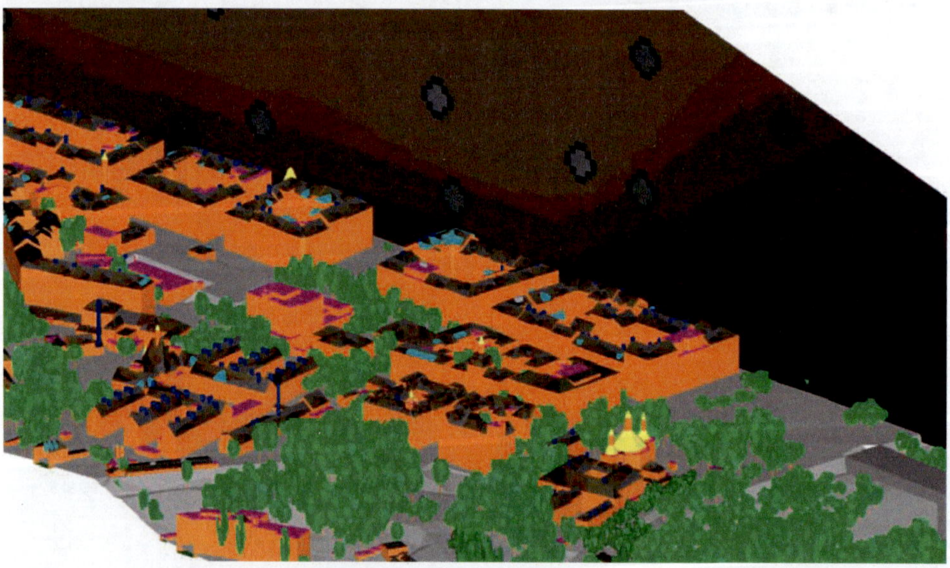

The detail view on the street canyon complemented by the image based on DIAL's data.

Figure 4. Processing of the DIAL-LIDAR data in the GIS and CAD environment.

8. CONCLUSION

A wide range of monitoring techniques and modeling procedures have been mentioned and shortly described in order to present the research activities focused on influence of the urban traffic on living conditions in urban areas. At present, the automotive traffic becomes the most important source of air pollution on the territory of large cities. Thus, the US EPA and the European Environmental Agency carry out many guidelines in the framework of various projects. The projects are shared by state agencies, research associations and universities from the whole world. Consecutively, the research outputs are implemented to the national legislation and guidelines [13, 14].

The environmental modeling supported by powerful computer tools and monitoring systems can predict dispersion and transport of pollutants. The environmental models used for regulatory applications are able to assist in decision-making processes of state authorities and local agencies. In order to improve the precision of predictions, there is still the need to develop new environmental models and to implement new monitoring techniques. A new research is represented by using data from high resolution satellite systems to identify surface objects and their properties in a more precise way. From this point of view, the remote sensing based on the DIAL-LIDARs represents the derived method that can bring a new insight into more detailed on-site air pollution estimates. The relationship between the urban morphology and the air pollution distribution can also help in the validation of numerical models and in a more realistic simulation in the wind tunnels [29, 30].

The attached case study focused on air pollution of the selected urban area in Prague introduces methodology implemented on the local level of the city Prague. In addition to the widely used techniques, spatial interpolations in the GIS environment based on data from the surface monitoring stations and the DIAL-LIDAR measurements are carried out to explore and visualize the distribution of the nitrogen oxides. However, detailed on-site air pollution estimates by the DIAL-LIDAR are not available for each road and time. But, in comparison with the estimates based on surface measurements and the standard modeling procedures, it can improve the prediction of air pollution in higher levels over the surface [41].

ACKNOWLEDGMENTS

The attached case study is partly based on the data created in the framework of the project GACR 205050898, which was generally focused on measurement and modeling of air pollution in urban areas and flat landscape. The authors wish to thank the Lidar company managed by A. Cerny, P. Berger, M. Strizik and P. Engst.

The input spatial data for building of the 3D scenes are originally provided by the Institute of Municipal Informatics of Prague. The map layers of emission sources (registered as REZZO 1 and REZZO 2) and the temporal measurements of the surface air pollution transferred from the Internet are administrated by the IT Department of the Prague City Hall.

The spatio-temporal data were processed by ESRI-ArcGIS and Leica Geosystems-ERDAS Imagine in the GIS Laboratory, Faculty of Natural Science, Charles University in Prague in the framework of the project AVCR 1ET400760405 supported by the Information Society, Academy of Sciences in the Czech Republic.

REFERENCES

[1] Freijer, J.I., Bloemen, H.J.T., Loos, S., Marra, M., Rombout, P.J.A., Steentjes, G.M. and Veen, M.P. (1998). Modelling exposure of the Dutch population to air pollution. *Journal of Hazardous Materials 61*, 107-114.

[2] Venkatram, A., Isakov, V., Thoma, E. and Baldauf, R. (2007). Analysis of air quality data near roadways using a dispersion model. *Atmospheric Environment 41*, 9481-9497.

[3] Skote, M., Sandberg, M., Westerberg, U., Claesson, L. and Johansson A.V. (2005). Numerical and experimental studies of wind environment in an urban morphology. *Atmpospheric Environment 39*, 6147-6158.

[4] Wang, L., Parker, D.B., Parnell, C.B., Lacey, R.E. and Shaw, B.W. (2006). Comparison of CALPUFF and ISCST3 models for predicting downwind odor and source emission rates. *Atmospheric Environment 40*, 4663-4669.

[5] Song, Y., Zhang, M. and Cai, X. (2006). PM10 modeling of Beijing in the winter. *Atmospheric Environment 40*, 4126-4136.

[6] Souto, M.J., Souto, J.A., Munuzuri, V.P., Casares, J.J. and Bermudez, J.L. (2001). A comparision of operational Lagrangian particle and adaptive puff models for plume dispersion forecastting. *Atmospheric Environment 35*, 2349-2360.

[7] Hayes, E.T., Curran, T.P. and Dodd, V.A. (2006). A dispersion modeling approach to determine the odour impact of intensive poultry production units in Ireland. *Bioresource Technology 97*, 1773-1779.

[8] Gromke, C. and Ruck, B. (2007). Influence of trees on the dispersion of pollutants in an urban street canyon-Experimental investigation of the flow and concentration field. *Atmospheric Environment 41*, 3287-3302.

[9] Meroney, R.N. and Bienkiewicz, B. (1997). *Computational Wind Engineering 2*. Amsterdam: Elsevier.

[10] Fujino, T., Asaeda, T. and Ca, V.T. (1999). Numerical analyses of urban thermal environment in a basin climate-application of a k-ε model to complex terrain. *Journal of Wind Engineering and Industrial Aerodynamics 81*, 159-169.

[11] Cressie, N.A.C. (1993). *Statistics for Spatial Data*. New York: Wiley.

[12] Johnston, K., Ver Hoef, J.M., Krivoruchko, K. and Lucas, N. (2001). *Using ArcGIS Geostatistical Analyst*. Redlands, California: ESRI Press.

[13] Seika, M., Metz, N. and Harrison, R.M. (1996). Characteristics of urban and state emission inventories-comparison from Europe and the United States. *The Science of the Total Environment 189/190*, 221-234.

[14] EMEP/CORINAIR Emission Inventory Guidebook-2007. Technical report No 16/2007. Copenhagen: European Environmental Agency.

[15] Salt, J.E. and Moran, A. (1997). International greenhouse gas inventory systems: A comparison between CORINAIR and IPCC methodologies in the EU. *Global Environmental Change 7(4)*, 317-336.

[16] Borrego, C., Tchepel, O., Barros, N. and Miranda, A.I. (2000). Impact of road traffic emissions on air quality of the Lisbon region. *Atmospheric Environment 34*, 4683-4690.

[17] Thorpea, A.J., Harrisona, R.M., Boulterb, P.G. and McCrae, I.S. (2007). Estimation of particle resuspension source strength on a major London Road. *Atmospheric Environment 41*, 8007–8020.

[18] Monforti, F. and Pederzoli, A. (2005). THOSCANE: a tool to detail CORINAIR emission inventories. *Environmental Modelling & Software 20(5)*, 505-508.

[19] Costa, M. and Baldasano, J.M. (1995). Development of a source emission model for atmospheric pollutants in the Barcelona area. *Atmospheric Environment 30(2)*, 309-318.

[20] Klimont, Z., Amann, M., Cofala, J., Gyárfáš, F., Klaassen, G. and Schöpp, W. (1994). An emission inventory for the central European initiative 1988. *Atmospheric Environment 28(2)*, 235-246.

[21] Paine, D.P. and Kiser, J.D. (2003). *Aerial Photography and Image Interpretation.* New York: Wiley.

[22] Lassueur, T., Joost, S. and Randin, C.F. (2006). Very high resolution digital elevation models: Do they improve models of plant species distribution? *Ecological Modelling 198*, 139-153.

[23] Molero, F. and Jaque, F. (1999). The laser as a tool in environmental problems. *Optical Materials 13*, 167-173.

[24] Gondal, M.A. and Mastromarino, J. (2000). Lidar system for remote environmental studies. *Talanta 53*, 147-154.

[25] Dell'Aglio, M., Kholodnykh, A., Lassandro, R. and De Pascale, O. (2002). Development of a Ti: Sapphire DIAL system for pollutant monitoring and meteorological application. *Optics and Lasers in Engineering 37*, 233-244.

[26] Fiorani, L. and Durieux, E. (2001). Comparison among error calculation in differential absorption lidar measurements. *Optics & Laser Technology 33*, 371-377.

[27] Kim, S.W., Yoon, C.S., Won, J.G. and Choi, S.C. (2007). Geound-based remote sensing measurements of aerosol and ozone in an urban area: A case study of mixing height evolution and its effect on ground-level ozone concentrations. *Atmospheric Environment 41*, 7069-7081.

[28] Zwoździak, J., Zwoździak, A., Sówka, I., Ernst, K., Stacewicz, T., Szymański, A., Chudzyński, S., Czyżewski, A., Skubiszak, W. and Stelmaszczyk, K. (2001). Some results on ozone vertical distribution in atmospheric boundary layer from LIDAR and surface measurements over the Kamienczyk Valley, Poland. *Atmospheric Research 58*, 55-70.

[29] Matejicek, L. (2005). Spatial modelling of air pollution in urban areas with GIS: a case study on integrated database development. *Advances in Geosciences 4*, 63-68.

[30] Matejicek, L., Engst, P. and Janour, Z. (2006). A GIS-based approach to spatio-temporal analysis of environmental pollution in urban areas: A case study of Prague's environment extended by LIDAR data. *Ecological Modelling 199*, 261-277.

[31] Briggs, D.J., de Hoogh, C., Gulliver, J., Wills, J., Elliott, P., Kingham, S. and Smallbone, K. (2000). A regression-based method for mapping traffic-related air pollution: application and testing in four contrasting urban environments. *The Science of the Total Environment 253*, 151-167.

[32] Pujadas, M., Plaza, J., Terés, J., Artínano, B. and Millán, M. (2000). Passive remote sensing of nitrogen dioxide as a tool for tracking air pollution in urban areas: the Madrid urban plume, a case of study. *Atmospheric Environment 34*, 3041-3056.

[33] Maidment, D.R. (1996). Environmental Modeling within GIS. In M.F. Goodchild, L.T. Steyaert, B.O. Parks (Ed.), *GIS and Environmental Modeling: Progress and Research Issues* (pp. 315-323). Fort Collins: GIS World, Inc.

[34] Fedra, K. (1996). Distributed Models and Embedded GIS: Integration Strategies and Case Studies. In M.F. Goodchild, L.T. Steyaert, B.O. Parks (Ed.), *GIS and Environmental Modeling: Progress and Research Issues* (pp. 413-417). Fort Collins: GIS World, Inc.

[35] Maguire, D.J., Batty, M., and Goodchild, M.F. (2005). *GIS, Spatial Analysis and Modeling*. Redlands, California: ESRI Press.

[36] Mitchell, A. (2005). *The ESRI Guide to GIS Analysis. Volume 2: Spatial Measurements and Statistics*. Redlands, California: ESRI Press.

[37] Maguire, D.J., Batty, M., and Goodchild, M.F. (2005). *GIS, Spatial Analysis and Modeling*. Redlands, California: ESRI Press.

[38] Grünfeld, K. (2005). Integration spatio-temporal information in environmental monitoring data-a visualization approach applied to moss data. *Science of the Total Environment 347*, 1-20.

[39] MacDonald, A. (2001). *Building a Geodatabase*. Redlands, California: ESRI Press.

[40] Solc, J. and Pospisilova, J. (2007). Prague-Environment 2006. J.I.M. Liberec: Prague.

[41] Matejicek, L. (2006). Spatial estimations of air pollution in urban areas by LIDAR measurements. In: A., Voinov, A.J., Jakeman, A.E., Rizzoli (Ed.), *Proceedings of the iEMSs Third Biennial Meeting: "Summit on Environmental Modelling and Software"*. International Environmental Modelling and Software Society, Burlington, USA, July 2006.

In: Environmental Modeling with GIS
Editor: Lubos Matejicek, pp. 99-111

ISBN: 978-1-60876-363-4

Chapter 7

MODELING OF TRAFFIC-RELATED ENVIRONMENTAL POLLUTION IN THE GIS

***Lubos Matejicek*[1] *and Zbynek Janour*[2]**
[1]Institute for Environmental Studies, Charles University in Prague
Benatska 2, 128 01, Prague, Czech Republic
[2]Institute of Thermomechanics, Academy of Sciences
Dolejskova 5, 182 00, Prague, Czech Republic

ABSTRACT

The numerical models are based on dispersion modeling and statistical analysis. In case of dispersion modeling, the ISC-AEROMOD View is used for modeling multiple pollutants with the U.S. EPA modeling tool ISCST3. The Mobile View assists as an interface for the U.S. EPA MOBILE6 model that predict arterial street emissions focused on hydrocarbons, carbon monoxide, nitrogen oxides, carbon dioxide, particulate matter, and toxics from cars, motorcycles, light- and heavy-duty trucks under various conditions. The potential impacts of accidental releases are solved by SLAB View that complements the modeling tools by analysis of emissions from accidental releases of toxic gases. Analysis of urban traffic-induced noise pollution is assessed by U.S. FHWA-TNM tools. The GIS is finally used to serve as a common analysis framework for individual modeling tools. In order to display the numerical simulation outputs together with urban area map layers, numerical modeling based on U.S. EPA software tools is integrated into the GIS for spatial interpolations and spatial analysis. It assists to evaluate high levels of air pollution and noise pollution together with the thematic map layers of residential zones, business centers, schools, and hospitals. Finally, finding alternative routes can decreases air pollution and noise pollution in selected zones. As a case study, the city of Prague sample data set helps to demonstrate data processing and modeling of traffic-related environmental pollution. The ESRI's geodatabase is used for implementation a comprehensive information model and a transaction model in the GIS environment. It is also the common application logic used in ArcGIS for accessing and working with all spatial thematic data and simulation inputs/outputs. Spatial interpolation for prediction maps and probability maps complement the existing thematic map layers, which enable cell based modeling for spatial multi criteria decision analysis. The synthesis of environmental models and GISs creates a more complex base for environmental simulation that can support decision-making processes in a more straightforward way.

Keywords: *Road traffic; air pollution; noise assessment; spatial modeling; GIS.*

1. INTRODUCTION

The increasing traffic in growing urban areas amplifies certain problems. It affect human daily life, because the traffic related emissions and noise are today's leading cause of environmental pollution due to the overall success in reducing emissions of pollutants from other industrial sources in the past decades. Numerous studies have identified many associations between traffic-related air pollution merged with other negative phenomena (noise, traffic accidents, traffic jams), and adverse potential health effects indicated either by characterizing exposures to specific pollutants using measurements from a few ambient sites or by some estimates of traffic [1, 2]. Despite ignoring the local contribution of indoor sources, the local effect of residential ventilation and the accuracy to estimate personal exposures, home-specific outdoor measurements are often used for initial estimates. In lieu of using these measurements, which requires nearly labor-intensive steps as indoor monitoring, some factors can be generated from the geographic information system (GIS) [3, 4]. It includes using distance from roads, complex digital elevation models, land use, and population density in combination with central site monitoring data to derive estimates of ambient exposures. Questionnaire data focused on individual building characteristics (air conditioning usage, opening windows and local indoor emission sources) can be used to particularize local environmental pollution and potential health effects.

Many current studies deal with publicly available data from monitoring networks and questionnaire responses managed by state agencies and local authorities. Some of them provide directions how publicly available data can be utilized, in order to predict residential indoor exposures in the absence of measurements. For example, information on traffic applied in the GIS framework in combination with ambient monitoring data focused on nitrogen dioxide, fine particulate matter and elemental carbon is used as a substitute for home-specific outdoor measurements and consecutively as a particular estimate for indoor exposures of outdoor dominated pollutants. Thus, models based on regression analysis and Bayesian approaches are often used for predictions [5, 6, 7]. Another research is represented by using methods of numerical modeling [8] or physical modeling [9]. Numerical models based on partial differential equations can solve dynamic phenomena of pollutant dispersion in dependence on wind flows above complex terrain. In addition to case oriented numerical models dedicated to special mathematical tasks, many software tools are accessible from U.S. EPA in the framework of the given guidelines. Physical models, exploring the relation between three-dimensional morphology and windiness in wind tunnel experiments, are usually used for prediction of pollutant dispersion in the idealized city models.

A relatively new approach is represented by statistical tools and numerical models integrated in the GIS environment. The GIS offers many tools for exploratory spatial data analysis and management, which can support numerical models in a more efficient way [10].

Environmental pollution by noise brings health effects and behavioral problems. Noise pollution can cause hypertension, high stress levels, hearing loss and sleep disturbances. In case of animals, noise causes stress and increases risk of mortality. Mitigation of traffic noise is provided by noise barriers, limitation of vehicle speed and acceleration. Applying these strategies optimizes computer models based on physical rules and local topography [11]. Again, GISs offer a wide range of tools for data management and noise modeling support.

Other potential sources of environmental pollution are impacts of accidental releases. They can cause releases of toxic gases that may expose residential zones, schools and hospitals in urban areas. Thus, the potential accidents at selected sites of crossroads and arterial roads are also included in modeling of traffic-related environmental pollution [12].

2. Traffic-Related Environmental Pollution

Environmental pollution originated from traffic comes from a wide variety of individual moving sources. Thus, in order to provide computer simulations, the crossroads are mostly identified as the point sources, the highways and city streets as the line sources, and the parking places or arterial roads as the area sources. In case of this paper, the environmental pollution includes air pollution, air pollution from potential car accidents and noise.

The air pollutants emitted directly into the atmosphere represent primary pollutants. The air pollutants formed in the air as a result of chemical reactions are known as secondary pollutants. For example, carbon monoxide (CO) and sulfur dioxide (SO_2) are primary pollutants, while ozone is a secondary pollutant. Nitrogen dioxide (NO_2) and some particulate matter represent both primary and secondary pollutants, because they are emitted directly into the atmosphere, and formed from other pollutants. In case of traffic emissions, some nitrogen dioxide is emitted directly from vehicle exhaust, but most is formed by oxidation of nitric oxide (NO) in the atmosphere. Similarly, fine particular matter is emitted directly by moving vehicles as well as formed in the air. In case of primary pollutants, the ambient concentrations indicate approximately proportional relationship to emissions. In case of secondary pollutants, the relationship is more complex. For example, reducing local emissions of nitrogen oxides can lead to an increase in local ozone concentration and other consequent reactions [13].

The emissions of carbon dioxide (CO_2), SO_2 and black smoke can be estimated from petrol and diesel consumption. Air pollution caused by SO_2 usually varies in dependence on the sulfur content of the fuels. Black smoke emissions are estimated in accordance with soiling factors for different fuels and vehicle types. The estimates of emissions of NO_x, CO and non-methane volatile organic compounds (NMVOCs) are often based on vehicle performance (emissions from cold starts, hot engines and evaporative losses) rather than fuel consumption. Thus, the emissions from individual motor vehicles mainly depend on the type of vehicle, the fuel used, and the vehicle performance that depends on configuration of the traffic network, especially in urban areas. In order to obtain required information, complex exploration and measurements of a number of various vehicles together with vehicle kilometers driven in roads are needed. In case of using the GIS, its spatial database can support mapping of traffic networks, classification of road types (urban, rural single or dual carriageway, motorway), mapping of vehicles categories (petrol or diesel cars, light goods or heavy duty vehicles, buses or coaches, and motorcycles) in dependence on spatial location. In addition to these categories, the vehicle type can be split according to the required regulation in the last years. The emission factors are mostly based on on-road measurements of emissions. Due to large variations from vehicle to vehicle, the emission factors used for calculations are represented by means of the measured emissions. Besides these uncertainties, a number of other implicit assumptions are needed like, for example, impacts of load, road gradient, etc. Particulate matters like PM_{10} need slightly different estimates in dependence on the type of vehicles and catalyst technology.

The prediction of future emissions based on specified factors also requires estimates of economic activity and demographic influence together with repeated terrain measurements. Information about emission forecasting models, annual reports and methodological recommendation are regularly published by U.S. EPA and by the European Environmental Agency (EMEP/CORINAIR Emission Inventory Guidebooks).

Traffic noise represents other part of environmental pollution. Various methodologies are used for risk assessment in dependence on recommendations of government authorities. Traffic noise prediction models mostly assist in simulation of sound pressure levels, specified in terms of the equivalent continuous level (L_{eq}) over a chosen period and under interrupted or varying flow conditions. The early models were focused on prediction of linear levels whereas the later models on prediction of A-weighted levels. For setting of the noise sources, single point sources, short line sources or multi point/line sources are taken into calculations, optionally, some with different spectra. Among many prediction models and case oriented studies [14, 15, 16], there is a few modeling tools recommended by national regulations (the FHWA model in the US, the CRTN model in the UK, the RLS90 model in Germany, the OAL model in Austria, the EMPA model in Switzerland, the ASJ model in Japan).

3. Description of Software Tools Used for Modeling of Traffic-Related Environmental Pollution

In the framework of the attached case study a number of software tools are used for modeling of traffic-related environmental pollutants in the selected urban area. In order to merge all the needed data for risk assessment, the GIS is utilized for data management, spatial analyses and visualization. Sharing data between the GIS and the individual modeling software tools is mostly through the shared data files.

Industrial Source Complex Short Term (ISCST), originally developed in the 1970's, is the US EPA's regulatory tool based on a steady-state Gaussian plume algorithm. The current version (ISCST3) contains many enhancements that include an improved area source algorithm, the complex terrain screening algorithms, a revised dry and wet deposition tools, and many revisions intended for air toxics applications. The ISCST3 is applicable for estimating ambient impacts from point, area and volume emission sources out to a distance of approximately 50 kilometers. In spite of that the ISCST3 is primarily dedicated to the dispersion modeling of the stationary sources, it can by applied together with other modeling tools for dominant traffic-related sources (sites with continuous traffic jams, larger parking places, and arterial roads during rush hours) [17, 18, 19].

The MOBILE6 includes a revised emission factor model for estimation of gram per mile emissions of hydrocarbons (HC), carbon monoxide (CO), nitrogen oxides (NO_x), carbon dioxide (CO_2), particulate matter (PM), and toxics from cars, trucks, and motorcycles under various conditions affecting in-use emission levels. Thus, the ambient temperatures and average traffic speeds can be specified by users to develop emission inventories and control strategies. Due to the new vehicles types and progress in traffic systems, the emission factor model has been repeatedly improved and revised since its initial version. Many case studies bring new ideas into research of the traffic-related emission estimates by adopting new technology [20, 21] and regional driving differences [22].

The FHWA Traffic Noise Model (FHWA TNM) for highway traffic noise prediction and analysis was initially released in 1998. Actual version, FHWA TNM 2.5, contains many improvements and revisions. As sources of noise, it includes noise emission levels for automobiles, medium tracks, heavy trucks, buses, and motorcycles. Noise emission levels consist of A-weighted sound levels (one-third octave-band spectra), and subsource-height strengths for a few pavement types (dense-graded asphaltic concrete, Portland cement concrete, open-graded asphaltic concrete, and composite pavement types consisting of previous listed types). It includes full-throttle noise emission levels for vehicles on upgrades, and vehicles accelerating away from stop signs, toll booths, traffic signals and on-ramp start points. Upcoming case studies are mostly focused on evaluation the effects of noise barriers, and assessing the model sensitivity to parameter variations [23, 24].

The computer model focused on the potential impacts of accidental releases, SLAB, complements the list of the selected models for presented complex modeling of traffic-related environmental pollution. The SLAB model solves the atmospheric dispersion of denser-than-air releases over flat. All the source input conditions have to be determined externally by approximate estimates or terrain measurements. A few release types are treated. They include continuous release, finite duration release, instantaneous release, ground-level evaporating pool, elevated horizontal jet, stack or elevated vertical jet and instantaneous volume source. The SLAB model was developed in 1990 by D.L. Ermak of the Lawrence Livermore National Laboratory with support from U.S. Department of Energy, USAF Engineering and Services Center, and the American Petroleum Institute [25].

A wide range of various data based on terrain measurements, existing projects, and annual reports needs to be managed and integrated together with modeling tools, in order to enable the exploration of traffic related environmental pollution in dependence on spatial scales, time scales, and other attributes. Thus, the GISs are used for most of the tasks. The described U.S. EPA modeling tools (ISCST3, MOBILE6, FHWA-TNM 2.5, and SLAB) complement the risk assessment information system as the case oriented modeling tools. Integration of the GIS and spatially distributed environmental models is based on pre- and postprocessor linkage through shared data files. Despite building models as analytical functions into the fully functional GIS, the U.S. EPA modeling tools and the GIS form standalone software systems. The preprocessed GIS data are imported into U.S. EPA models, and simulation outputs are backward exported into the GIS to provide spatial analysis and visualization. Thus, the GIS project is finally used to serve as a common analysis framework for individual U.S. EPA modeling tools. In addition to standard GIS functionality (entering, storing, retrieving, transforming, measuring, combining, subsetting, and displaying spatial data that are registered to a common coordinate system) [26], managing 3D data for complex digital terrain models (DTMs extended by buildings, barriers, and other surface objects) and advanced spatial interpolation algorithms are needed to be carry out for final display and visualization. To accomplish all the needs, the ESRI's geodatabase offers to implement a comprehensive information model and a transaction model in the GIS environment. Thus, the ArcGIS tools can assess and work with all spatial thematic data and simulation inputs/outputs. The GIS can also efficiently handle data from remote sensing and global positioning systems (GPSs). Thus, the aerial images and terrain observations complement the existing map layers in the framework of the final visualization. The satellite images can extend the map layers by data from other infrared spectral bands. GPSs, frequently used for car navigation, extend existing map layers by location of terrain measurements.

4. A Case Study for Modeling of Traffic-Related Environmental Pollution

In order to demonstrate modeling of the traffic-related environmental pollution, the GIS project is created to integrate dispersion of pollutants, noise assessment and potential accident releases together with terrain measurements, aerial images, satellite images, DTMs, annually reported data and other existing spatial data. The data processing and modeling in the framework of the GIS project is illustrated in Figure 1.

The upper part of the schema includes the main environmental data inputs. In case of air pollution, the meteorological data contain local estimates of the wind speed and the wind direction. In addition to importing the DTM into the modeling tools (ISC AEROMOD View Terrain Processor, FHWA-TNM and SLAB), the simplified shapes of buildings are used for US EPA BPIP Model, in order to effectively and quickly complete the building downwash analysis in the framework of running the ISCST3. The Computer Aided Design (CAD) files include the DTM complemented by surface objects. Traffic related data describe traffic intensity and vehicle types for setting of the modeling tools based on the MOBILE6 and the FHWA-TNM. The next phase includes pre-processing of the input data into the data formats for inputs of the selected models (ISCST3, MOBILE6, FHWA_TNM, and SLAB). In case of the ISCST3, the PC-RAMMET is used for pre-processing of meteorological data and the BPIP for adopting of spatial data (DTM and buildings). While the described data processing can be carried out in the GIS (with exception of the PC-RAMMET, and the BPIP), the models solved by ISCST3, MOBILE6, FHWA-TNM and SLAB are implemented in the standalone software tools. In spite of that they also include display functions, the simulation outputs are imported backward into the GIS project that manages a wide range of spatial analyses, and provide more complex outputs for risk assessment. In addition to the spatial data management, the spatial interpolations by deterministic techniques (Inverse Distance Weighted-IDW, and polynomial interpolations) and geostatistical methods (kriging and cokriging) assist for estimations of air pollution concentrations in the neighborhood of the sites predicted by models, and existing monitoring stations [27, 28]. The GIS also contains many spatial exploratory statistical tools such as histograms for checking normal distribution, quantile plots, trend analysis and semivariograms. The normal distribution of input data is required for geostatistical interpolation methods that produce not only prediction surfaces but also uncertainty surfaces, give an indication of prediction quality and can generate probability and quantile output maps depending on spatial modeling needs. Unsuitable sites can be located by combining layers focused on air pollution and noise assessment. After the deriving the air pollution layers and the noise level layer to a common scale, the cell based modeling can assist in calculation the output map for complex risk assessment.

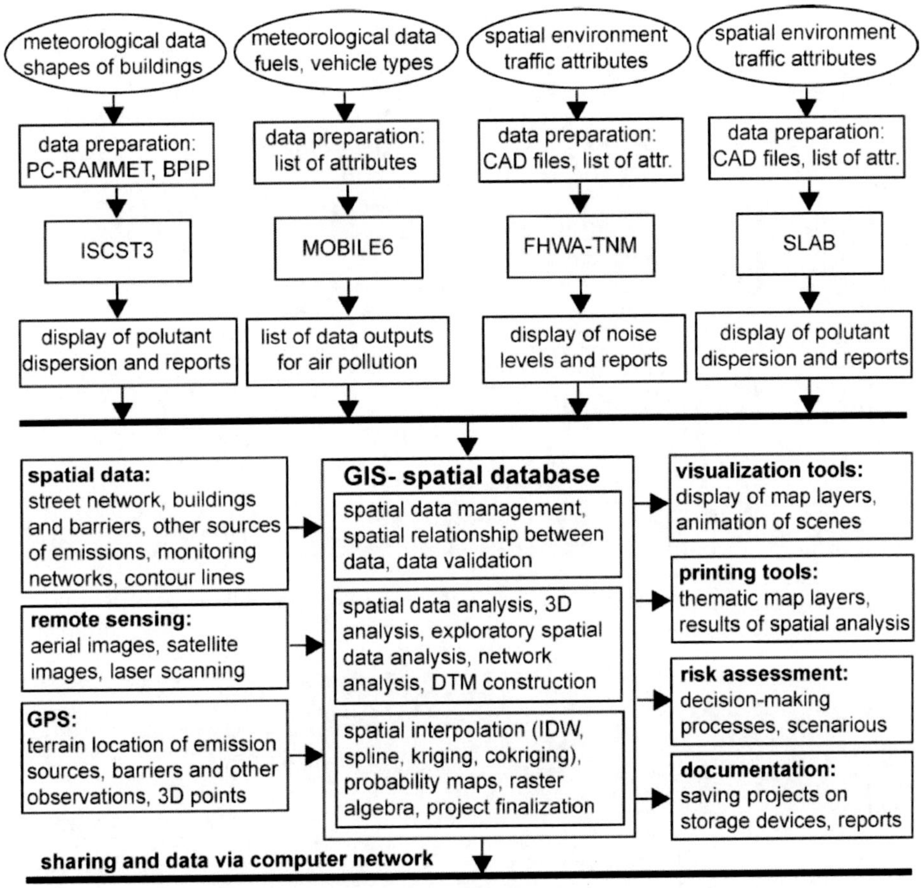

Figure 1. The flow diagram for data processing and modeling of the traffic-related environmental pollution in the GIS environment.

4.1 Mapping of the emission sources

As an example, the urban site in Prague, the Czech Republic, characterized by high population density is selected for modeling of traffic-related environmental pollution, Figure 2. The accumulation of traffic in the last decade has indicated an increasing pressure in the air quality and noise. The more detailed view of the main crossroad from the aerial image is in Figure 3. The image was not taken during the rush hours, which causes a small percentage of the vehicles on the roads. The significant spatial and temporal variability of traffic during the day also results in variability of traffic-related environmental pollution. A number of environmental studies have raised the question of how representative the site and time period of air quality actually can be in comparison with other sites and different time periods [29]. In order to provide validation of simulation outputs, data from surface monitoring stations and local mobile DIAL-LIDAR measurements can be used for the selected site [30].

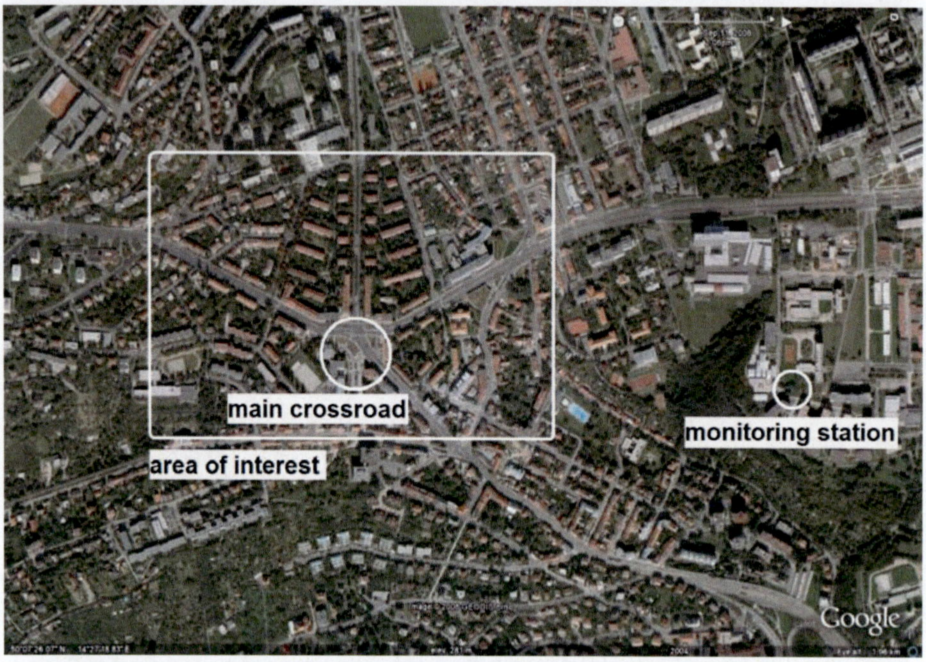

Figure 2. The selected site for modeling of traffic-related environmental pollution with a main crossroad in the middle part and arterial roads (Google Earth 2008, Image © GEODIS, Brno).

Figure 3. Simulation results of the ammonia horizontal jet release caused by a traffic accident at the main crossroad displayed together with the aerial images in the GIS environment.

In case of the presented case study, estimates of emissions from the motor vehicles are based on the mobile sources emission factor model implemented in the U.S. EPA MOBILE6. The setting of the parameters reflects the rush hours, in order to explore scenarios focused on expected higher levels of air pollution and noise. The traffic volumes, average vehicle speeds, humidity, and meteorological parameters are based on terrain observations and regular reports.

4.2. Simulation of traffic-related environmental pollution

As an example, the ammonia horizontal jet release caused by the traffic accident at the main crossroad is simulated and displayed together with other spatial data in the GIS environment, Figure 3. The SLAB model simulates the atmospheric dispersion in accordance to the spill source type, source properties, spill parameters, local field parameters and meteorological conditions. The spread of a plume caused by an accidental release of a chemical is drifted above the neighbor residential zone. Line contours give an approximate concentration levels.

Simulation of atmospheric dispersion originated from vehicles is demonstrated by the ISCST3 in the framework of the AEROMOD View. As an example, the simulation results based on a steady-state Gaussian plume algorithm are used for the approximate spatial prediction of NO_x in Figure 4.

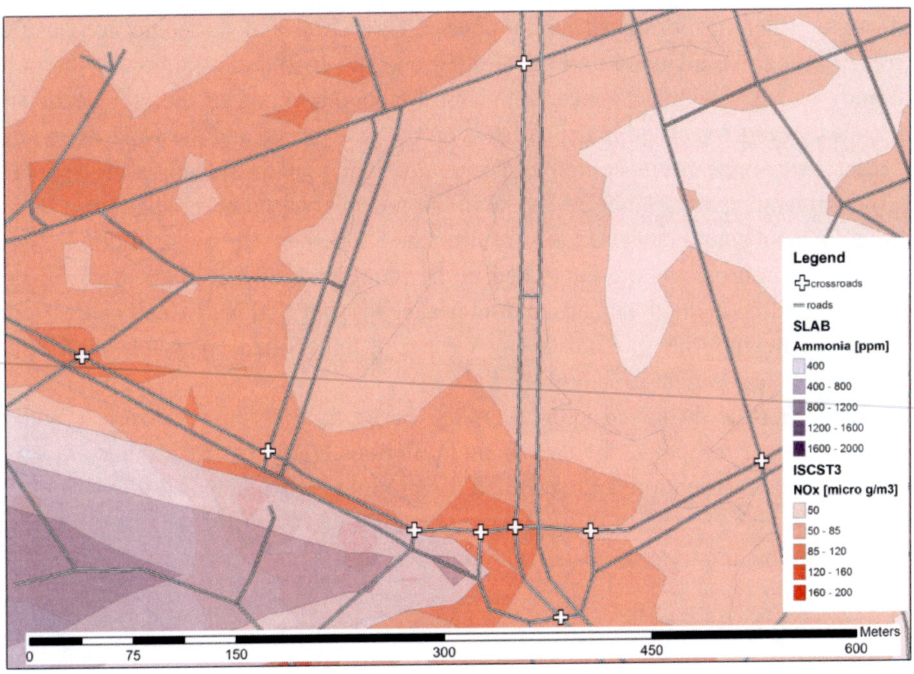

Figure 4. Simulation results of the ammonia horizontal jet release together with the traffic related air pollution (NO_x) displayed in the GIS environment.

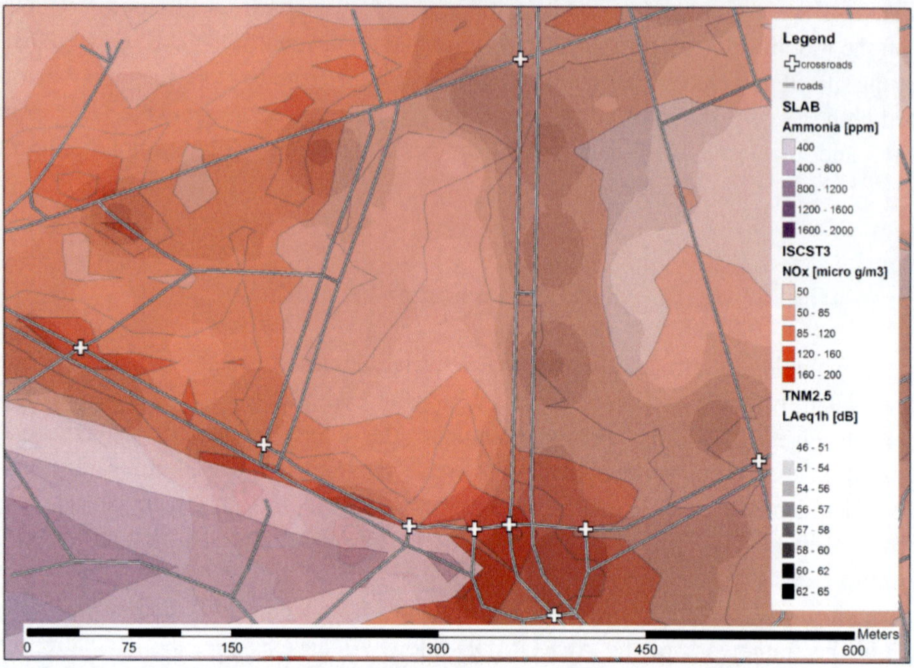

Figure 5. Simulation results of the ammonia horizontal jet release together with the traffic related air pollution (NO_x), and noise assessment displayed in the GIS environment.

But, many other traffic-related compounds (CO, VOCs, and dust emissions caused by heavy transport) can be included into the simulation. The spatial distribution of the pollutant concentration shows the higher concentration in the neighborhood of the crossroads and the arterial roads that can be shifted in dependence on the wind speed and the wind direction.

In order to provide more complex risk assessment, the traffic noise model (FHWA-TNM2.5) complements used models. The simulation results based on prediction of hourly A-weighted equivalent sound level (L_{Aeq1h}) are illustrated together with air pollution in Figure 5. Again, the simulation outputs are integrated in the framework of the GIS project. The higher noise levels are at the main crossroads and arterial roads, but optionally, can be decreased by barriers in their neighborhood. Considering to sharing other map layers, the GIS approach offers a more efficient way in design of the barriers.

In spite of that the displayed map layers are based on the dense network of individual receptors or grids of receptors, the spatial interpolations (IDW or the ordinary kriging, for normal input data distributions) are needed to produce continuous surfaces of pollutant concentrations or noise levels. In this case, the GIS provides a comprehensive set of tools for creating surfaces that are used for visualization and analysis.

Conclusion

In spite of that many environmental studies focused on air pollution, noise assessment and other environmental phenomena have been carried out [31, 32, 33], there is still need to develop research tools for complex risk assessment. Different techniques of terrain measurements, different methodology of data processing, and finally, various decision-making processes have created a wide range of individual procedures in dependency on spatial and temporal resolution of the areas of interest. In order to unify some common procedures of data processing, the GISs can be used as a framework for running and development of powerful, efficient and complex tools in air pollution exploration and modeling [34, 35, 36]. In addition to spatial database management, many extensions focused on network analysis, spatial analysis, 3D spatial modeling and geostatistical methods can support modeling of the traffic-related environmental pollution. The used case oriented modeling tools for the accidental releases, the pollutants dispersion, and the noise assessment have shown that new ways of integration are needed. In case of environmental data, the unified data storage complemented by spatial information from GPS or remote sensing can be realized by the spatial database in the GIS environment. It is designed to offer best available knowledge to bear on environmental planning and policy making, reach a broad audience, be easy to use and understand and help to explore a wide range of options. Thus, the close integration of environmental models into the GIS will help to explore the traffic-related environmental pollution in a more sophisticated way, and will reach a broad audience.

ACKNOWLEDGMENTS

The input spatial data included into the case study are originally provided by the Institute of Municipal Informatics of Prague. The map layers of emission sources and the temporal measurements of the surface air pollution transferred from the Internet and from the annual reports are administrated by the IT Department of the Prague City Hall.

The spatio-temporal data were processed by ESRI-ArcGIS and Leica Geosystems-ERDAS Imagine in the GIS Laboratory, Faculty of Natural Science, Charles University in Prague in the framework of the project AVCR 1ET400760405 supported by the Information Society, Academy of Sciences in the Czech Republic.

REFERENCES

[1] Seika, M., Metz, N. and Harrison, R.M. (1996). Characteristics of urban and state emission inventories-comparison from Europe and the United States. *The Science of the Total Environment 189/190*, 221-234.

[2] Salt, J.E. and Moran, A. (1997). International greenhouse gas inventory systems: A comparison between CORINAIR and IPCC methodologies in the EU. *Global Environmental Change 7(4)*, 317-336.

[3] Maguire, D.J., Batty, M. and Goodchild, M.F. (2005). *GIS, Spatial Analysis and Modeling*. Redlands, California: ESRI Press.

[4] Mitchell, A. (2005). *The ESRI Guide to GIS Analysis. Volume 2: Spatial Measurements and Statistics*. Redlands, California: ESRI Press.

[5] Brigs, D.J., de Hoogh, C., Gulliver, J., Wills, J., Elliott, P., Kingham, S. and Smallbone, K. (2000). A regression-based method for mapping traffic-related air pollution:

application and testing in four contrasting urban environments. *The Science of the Total Environment 253,* 151-167.

[6] Hochadel, M., Heinrich, J., Gehring, U., Morgenstern, V., Kuhlbusch, T., Link E., Wichmann, H.E. and Krämer, U. (2006). Predicting long-term average concentration of traffic-related air pollutants using GIS-based information. *Atmospheric Environment 40*, 542-553.

[7] Baxter, L.K., Clougherty, J.E., Paciorek, J., Wright, R.J. and Levy, J.I. (2007). Predicting residential indoor concentrations of nitrogen dioxide, fine particulate matter, and elemental carbon using questionnaire and geographic information system based data. *Atmospheric Environment 41*, 6561-6571.

[8] Venkatram, A., Isakov, V., Thoma, E. and Baldauf, R. (2007). Analysis of air quality data near roadways using a dispersion model. *Atmospheric Environment 41*, 9481-9497.

[9] Skote, M., Sandberg, M., Westerberg, U., Claesson, L. and Johansson A.V. (2005). Numerical and experimental studies of wind environment in an urban morphology. *Atmpospheric Environment 39*, 6147-6158.

[10] Maidment, D.R. (1996). Environmental Modeling within GIS. In M.F. Goodchild, L.T. Steyaert, B.O. Parks (Ed.), *GIS and Environmental Modeling: Progress and Research Issues* (pp. 315-323). Fort Collins: GIS World, Inc.

[11] Steele, C. (2001). A critical review of some traffic noise prediction models. *Applied Acoustics 62*, 271-287.

[12] Jaeger, J.A.G., Bowman, J., Brennan, J., Fahrig, L., Bert, D., Bouchard, J., Charbonneau, N., Frank, K., Gruber, B. and von Toschanowitz, K.T. (2005). Predicting when animal populations are at risk from roads: an interactive model of road avoidance behavior. *Ecological Modelling 185*, 329-348.

[13] Holman, C. (1999). Sources of Air Pollution. In S. Holgate, H. Koren, J. Samet, R. Maynard (Ed.), *Air Pollution and Health* (pp. 115-148). Amsterdam: Elsevier.

[14] Sopasakis, A. (2004). Stochastic noise approach to traffic flow modeling. Physica A 342, 741-754.

[15] El-Fadel, M., Shazbak, S., Baaj, M.H. and Saliby, E. (2002). Parametric sensitivity analysis of noise impact of multihighways in urban areas. *Environmental Impact Assessment Review 22*, 145-162.

[16] Pamanikabud, P. and Vivitjinda, P. (2002). Noise prediction for highways in Thailand. *Transportation Research Part D 7*, 441-449.

[17] Wang, L., Parker, D.B., Parnell, C.B., Lacey, R.E. and Shaw, B.W. (2006). Comparison of CALPUFF and ISCST3 models for predicting downwind odor and source emission rates. *Atmospheric Environment 40*, 4663-4669.

[18] Lorber, M., Eschenroeder, A. and Robinson, R. (2000). Testing the USA EPA's ISCST-Version 3 model on dioxins: a comparison of predicted and observed air and soil concentrations. *Atmospheric Environment 34*, 3995-4010.

[19] Hao, J., He, D., Wu, Y., Fu, L. and He, K. (2000). A study of the emission and concentration distribution of vehicular pollutants in the urban area of Beijing. *Atmospheric Environment 34*, 453-465.

[20] Pokharel, S.S., Bishop, G.A. and Stedman, D.H. (2002). An on-road motor vehicle emissions inventory for Denver: an efficient alternative to modeling. *Atmospheric Environment 36*, 5177-5184.

[21] Kuhns, H.D., Mazzoleni, C., Moosmüller, H., Nikolic, D., Keislar, R.E., Barber, P.W., Li, Z., Etyemezian, V. and Watson, J.G. (2004). Remote sensing of PM, NO, CO and HC emission factors for on-road gasoline and diesel engine vehicles in Las Vegas, NV. *Science of the Total Environment 322*, 123-137.

[22] Lin, J. and Niemeier, D.A. (2003). Regional driving characteristics, regional driving cycles. *Transportation Research Part D 8*, 361-381.

[23] El-Fadel, M., Shazbak, S., Baaj, M.H. and Saliby E. (2002). Parametric sensitivity analysis of noise impact of multihighways in urban areas. *Environmental Impact Assessment Review 22*, 145-162.

[24] Li, B., Tao, S., Dawson, R.W., Cao, J. and Lam, K. (2002). A GIS based road traffic noise prediction model. *Applied Acoustics 63*, 679-691.

[25] Thé, J.L., Thé, C.L. and Johnson, M.A. (2004). Slab View. User's Guide. Ontario, Canada: Lakes Environmental.

[26] Johnston, C.A. (1998). *Geographic Information Systems in Ecology*. London: Blackwell.

[27] Cressie, N.A.C. (1993). *Statistics for Spatial Data*. New York: Wiley.

[28] Johnston, K., Ver Hoef, J.M., Krivoruchko, K. and Lucas, N. (2001). *Using ArcGIS Geostatistical Analyst*. Redlands, California: ESRI Press.

[29] Vardoulakis, S., Gonzalez-Flesca, N., Fisher, B.E.A. and Pericleous, K. (2005). Spatial variability of air pollution in the vicinity of a permanent monitoring station in central Paris. *Atmospheric Environment 39*, 2725-2736.

[30] Matejicek, L., Janour, Z. and Strizik, M. (2008). Spatial Modeling of Air Pollution Based on Traffic Emissions in Urban Areas. In S.E. Paterson, L.K. Allan (Ed.), *Road Traffic: Safety, Modeling, and Impacts* (in press). New York: Nova Science Publisher, Inc.

[31] Mower, B. (1998). A multiple source approach to acute human health risk assessment. *Waste Management 18*, 377-384.

[32] Borrego, C., Tchepel, O., Barros, N. and Miranda, A.I. (2000). Impact of road traffic emissions on air quality of the Lisbon region. *Atmospheric Environment 34*, 4683-4690.

[33] Borrego, C., Tchepel, O., Costa, A.M., Martins, H., Ferreira, J. and Miranda, A.I. (2006). Traffic-related particulate air pollution exposure in urban areas. *Atmospheric Environment 40*, 7205-7214.

[34] Matejicek, L. (2005). Spatial modelling of air pollution in urban areas with GIS: a case study on integrated database development. *Advances in Geosciences 4*, 63-68.

[35] Matejicek, L., Engst, P. and Janour, Z. (2006). A GIS-based approach to spatio-temporal analysis of environmental pollution in urban areas: A case study of Prague's environment extended by LIDAR data. *Ecological Modelling 199*, 261-277.

[36] Grünfeld, K. (2005). Integration spatio-temporal information in environmental monitoring data-a visualization approach applied to moss data. *Science of the Total Environment 347*, 1-20.

In: Environmental Modeling with GIS
Editor: Lubos Matejicek, pp. 113-127
ISBN: 978-1-60876-363-4

Chapter 8

ESTIMATING THE ENVIRONMENTAL EFFECTS ON RESIDENTIAL PROPERTY VALUE WITH GIS

Lubos Matejicek
Institute for Environmental Studies, Charles University in Prague, Faculty of Natural Science, Benatska 2, 128 01, Prague, Czech Republic

ABSTRACT

Ongoing research to develop a new generation of decision-making tools for estimating the environmental effects on residential property value has significantly increased the demand for land surface data, information on the state of living environment, and the corresponding need for the advanced use of geographic information systems (GIS). In order to provide the foundation for price estimates, all the existing data focused on building and environment are integrated in the framework of a GIS spatial database. Traditional methods mostly emphasize the relationship between the effects of accessibility to central locations, and ignore location-specific attributes of housing. However, little has been done on high-rise, densely populated residential areas. Thus, the paper aims to investigate the neighboring and environmental characteristics of the selected site. A more advanced approach based on spatial analysis and modeling in the GIS environment is used to manage spatio-temporal data, to process aerial images and satellite images, to import measurements from GPS, to create digital terrain models, to analyze topography together with environmental data, and to visualize the results. The partial results based on the traveling time to the closest services and the origin-destination cost matrixes are derived from the network analysis. The landscape characteristics can be demonstrated on animated sequences showing the flight over a landscape model that is based on the digital terrain model with draped images over it to show the area. The living environment contains a few compartments: air pollution, water pollution, waste management, noise assessment, and monitoring of environmental impacts on population health. Air pollution is estimated by continuous map surfaces, predicting the values of pollutants concentration for the selected site in dependence on the sample points at the air quality monitoring stations. Water pollution covers surface water pollution, drinking water supply and quality, waste water, accidental contaminant spills, and optionally, flood control measures. The waste compartment is focused on the system of municipal waste management with sorting of reusable components of municipal waste, and hazardous chemicals information. Noise assessment covers road traffic noise and air traffic noise. The environmental impacts on population health come out from reports of the national institute of public health. In addition to the main compartments, other data can complement the partial inputs to the system (energy prices, public transport schema, locations of neighbor natural reservation sites and historical sites). In the framework of spatio-temporal analysis, the spatial weighting matrixes are used for prediction of the final rating. The final results are represented by the thematic map layers in the GIS project based on ArcGIS and its extensions. The attached case study shows a scenario in dependence on setting the weight parameters. Unexpected findings are caused by rapid changes in the dense living environment and slow conversion of the reality market in the selected site in Prague, the Czech Republic. For all that, the case study brings better understanding of how the residential site rating depends on various environmental attributes.

Keywords: Building; living environment; spatial modeling; GIS.

1. Introduction

Throughout history, the residential environment affects our health and well-being and we, through our activities, reshape the residential environment. This environment is interconnected to natural, social, and economic processes that run on the local, regional, and global scales. But, different neighborhoods or communities have different demographic characteristics, because of the economic and social processes that structure age, and race-ethnicity of the population.

The residence represents an important activity site in the environment where a person spends time. In addition to residence sites, the activity space is larger and depends on other activity sites like workplaces, stores, recreational areas, and the pathways traveled to and from the residential site. The size and shape of an individual person vary depending on its profession, the modes of transportation available and the geographical locations where all the activities take place. The residential activity sites together with others represent the zones where the individuals can be exposed to risk or resources, which affect the residential property value. It can be documented by disease mapping that has made contributions to public health and epidemiology for centuries. A common practice in mapping the residential property value by health events has been to calculate health event rates for political or administrative units like towns or tracts, because population and health outcome data are mostly reported for these areas. An obstacle with this approach is evident in arbitrary partitioning the underlying distribution of population and health events. Also, the risk factors, like population at risk, are not located at individual residential sites. Selected contaminants like air pollution, water pollution, waste, noise and biological agents are present in our living environment in various local and global scales. The temporal and spatial patterns are needed to model the distribution of known or potential risk factors.

2. GIS and Associated Systems

Geographic Information Systems (GISs) are being used in pollution environmental studies to model pollution dispersion and its effects on residential areas. The ability to estimate the environmental effects on residential property values is important for decision-making processes in order to start health public intervention activities [1]. Spatio-temporal modeling, in the framework of GIS, occurs whenever operations attempt to emulate geographic processes in the real world, at selected points in time or over extended periods. In the past, many other modeling applications have been coupled with GIS for high performance in dynamic modeling, remote sensing calculations and database management. But with the increasing power of GIS hardware and software, it is now possible to reconsider this relationship and to develop multifunctional tools [2]. Thus, the spatio-temporal modeling is trying to integrate several different contexts in the GIS world. There are two particularly important meanings.

First, spatial modeling by itself is focused on data models that include a set of expectations about data extended by spatial analysis. The data models are mostly represented by templates into which the data needed for particular applications are fitted together with topological rules and spatial analysis results. For example, the ESRI's geodatabase supports models of topologically integrated feature classes with extensions for complex networks, topologies, relationships among feature classes, and other object-oriented features. The geodatabase is implemented on standard relational databases complemented by spatial database extensions [3]. The way feature classes are often used in GIS, the features correspond to a group or class of real-world features, such as buildings, roads, railways, rivers, bus lines, dominant sources of environmental pollutions, and monitoring networks. In essence, GIS assists to develop the data models and to create map composition and visualization based on data models [4].

Second, temporal modeling deals with representation of processes that modify or transform some aspects of the Earth's surface through time. For example, contemporary environmental pollution forecasts are based on dynamic models of the particle dispersion in air, water and soil. Temporal models of vehicles behavior are used to predict traffic congestions. The strategies of integration spatially distributed environmental models into GIS range from simple linkage through shared data to building temporal models as analytical extensions into fully functional GISs. In case of shared data, two separate systems, the GIS and the simulation system, just exchange data through common files or a shared database [5].

GISs, computer based systems for the management and analysis of geographic data, integrate spatial data that originate from observation and measurement of the earth phenomena referenced to their locations on the earth's surface [6]. Whenever urban planners or environmental authorities use residential registration with address information, consider the location of pollution sources, or look at air quality and water quality reports from monitoring networks, they are working with geographic data. Thus, the GISs are considered as tools applicable to the integration and analysis of many different types of spatio-temporal data by people in different organizational setting asking various questions [7]. Thus, the GIS is related to other types of software that are focused on spatial data and spatial analysis [8].

2.1 GIS Mapping and Standard Digital Data Sources

The process and object views are expressed in the GIS by two main data structures: raster and vector. These data structures have different properties for storage and processing of spatial data. Majority of the recent GISs can manage these data structures, but many various data formats implemented in GISs require advanced data transformation methods [9].

In a raster system, each raster layer matches to a single attribute for a unit of space. For example, elevation or pollutant concentration at a particular place corresponds to a single attribute. Raster data are represented by an array of cells corresponding to particular places. The setting of the cell size affects the size of an object that can be discerned in a digital image, which determine the spatial resolution of the data. Landscape feature such as a single-family detached residence has to be discerned in high resolution raster images like the 10-meter spatial data, because the feature could be smaller than the raster cell size. On the other hand, low resolution is used to be provided for large geographical areas where the limited detail of landscape features is sufficient.

In a vector system, traditional GISs use layers that correspond to a single class of objects with the same dimensionality. Thus, the point layers can represent locations of residential buildings, sources of pollution, a network of air quality monitoring stations or locations of water quality measurements. The line layers represented by a sequence of straight-line segments interconnected with vertexes are used for traffic networks, surface water networks or line sources of pollution. The polygon layers formed by line boundaries can represent residential buildings in a more detailed view, different types of land-cover or land-use. In addition to the spatial location of vector objects, many attached attributes stored in tables contain other information. While the raster topology is based on regularly organized cells in grids, the vector topology has to reflect the neighborhood relationships among various types of vector objects.

In order to effectively store spatial data and their spatial relationship, a few advanced vector formats have been developed during last decades. For example, the ESRI's coverage model has been used for a few years. An extension of the coverage model is the ESRI's geodatabase that support complex networks, relationship among feature classes, and other object-oriented features. Both the models are also implemented on standard relational databases in the framework of the ArcSDE application server [10]. Thus, the individual environmental effects can be studied in dependence on location and time [11]. Mostly, they are represented by main environmental compartments like air pollution, water pollution, landscape, waste management, and noise assessment. The air pollution compartment contains layers focused on emissions, air quality monitoring, and predictions of pollutant concentrations. The water compartment can include layers with monitoring of surface water, drinking water, waste water, accidental contaminant spills, and flood control measures. The landscape compartment usually includes a wide range of thematic map layers focused on land cover, nature conservation, landscape protection, city greenery, and protected areas. The layers dealing with waste management include information about the system of municipal waste management. The noise assessment compartment include mapping of road and air traffic noise.

2.2 Remote Sensing and Field Mapping by GPS

Remote sensing deals with the acquisition of information about the entity without being in physical contact [12]. The data are collected from sensors on satellites or aircrafts. In case of Earth observations, two major groups of sensors can be used for collection of images in dependence on the source of electromagnetic spectrum.

Passive sensors detect natural energy that is emitted by the surface objects. Examples of passive remote sensors include airborne cameras or sensors carried by Landsat 7, Aster, Ikonos, and other existing satellites. Aerial images can assist in the high resolution mapping of residential zones and their surrounding areas. Satellite images offer to collect and transmit data from various parts of the electromagnetic spectrum (red-green-blue visible spectrum, near-middle-far infrared invisible spectrum), which in conjunction with aerial or ground-based sensing provides Earth science research with more information to monitor trends such as evolution of environmental and climatic changes.

Active sensors, on the other hand, emit energy in order to scan objects whereupon passive sensors then detect the radiation that is reflected from the target objects. Examples of these sensors represent 3D surface laser scanners or sensors carried by Radarsat. Radar/Lidar systems offer active remote sensing where the time delay between energy emission and its detection is measured in order to acquire the location and shape of the surface objects. It is often used for digital terrain mapping, which include the construction of a digital terrain model (DTM) [13]. In such a process, points are sampled by remote sensing sensors or by a global positioning system (GPS) from the terrain to be modeled with a certain observation density and accuracy. The terrain surface is then represented by the set of sample points that can be complemented by spatial interpolation if attributes on locations on the digital surface other than sample points is needed. In case of visualization, grid- and triangle-based modeling approaches are widely used than point-based approaches. For grid-base modeling, a grid network needs to be formed through spatial interpolation, if the original data are not in a grid form. For triangle-based modeling, a triangulation network is formed from point-based approaches through a triangulation procedure in order to create the triangular irregular network (TIN). Finally, the digital terrain model together with the draped satellite or the aerial images are included into other GIS layers for landscape analysis and visualization that are linked to the residential property value mapping [14, 15, 16].

3. Mapping of the Environmental Pollution

The success of environmental-related GIS studies depends on having access to accurate, timely, and compatible spatio-temporal data. Map layers focused on living environment mostly complement spatial data sets for a wide range of policy and planning issues. In case of mapping environmental pollution, their value extends well beyond the scope of the original projects for which they were created, and it increases as the data sets are used and updated. In order to use environmental data sets in GIS, the data sets must be captured and linked to the existing map layers in the framework of a foundation spatial database. In many cases, the digital data need to be imported into GIS from exiting information systems. Thus, the complex process of scanning, digitizing, entering field data and address matching can be reduced or even omitted.

The mapping of the environmental pollution together with other spatial data brings together four key elements: the spatio-temporal data stored in spatial database, the representations of that information on map layers, queries about the spatio-temporal data, and the analysis that creates new information and new map layers.

4. Spatial Analysis and Visualization

The environmental effects on residential property value affect agents that produce adverse health outcomes in humans. At residential zones, these agents can be physical (noise), chemical (ozone, nitrogen oxides, sulphur dioxide, particle fraction PM10) or biological (cryptosporidium) in nature. Human population encounters these agents by breathing, eating, and drinking, or coming into physical contact at built environment. The processes by which agents in the living environment produce an adverse health outcome can be modeled as hazard-exposure-outcome processes by risk assessment methods [17]. In the last decade, many models have been carried out in the field of noise impacts [18, 19], air pollution [20, 21, 22, 23], water pollution [24, 25], waste management [26] and land use/land cover changes [27].

In order to link spatio-temporal data drawn from many sources into the GIS, time and space must be the basis for data integration in a way that models hazard-exposure-outcome processes. Besides the integration of digital map layers focused on the location of residential buildings, hazardous agents need to be identified and their source locations and associated contamination fields must be estimated. In spite of that human experience is an important source of awareness of the harmful effects of many toxicants, there is still the need to carry out long term research in the framework of quantitative risk assessment that deals with the process of characterizing the health effects expected from exposure to an agent and estimating the probability of occurrence of health effects. Many case studies and reports are annually published by U.S. EPA. Subsequent legislation represents the major impetus for conducting risk assessment, but the responsibility for determining acceptable risk is in the range of state authorities and local agencies that implement the legislation. In this case, the methods of hazard identification where the GISs make the strong contribution include spatio-temporal association of adverse environmental residential effects [28, 29, 30].

5. A Case Study: Mapping of Environmental Pollution in Prague with Respect to Environmental Effects on Residential Property Value

The mapping of environmental pollution in the city Prague aimed to evaluate the application of spatio-temporal analysis with estimating the environmental effects on residential property value. The project was created in the framework of ESRI's ArcGIS and the spatio-temporal data were mostly saved in ESRI's geodatabase, Figure 1.

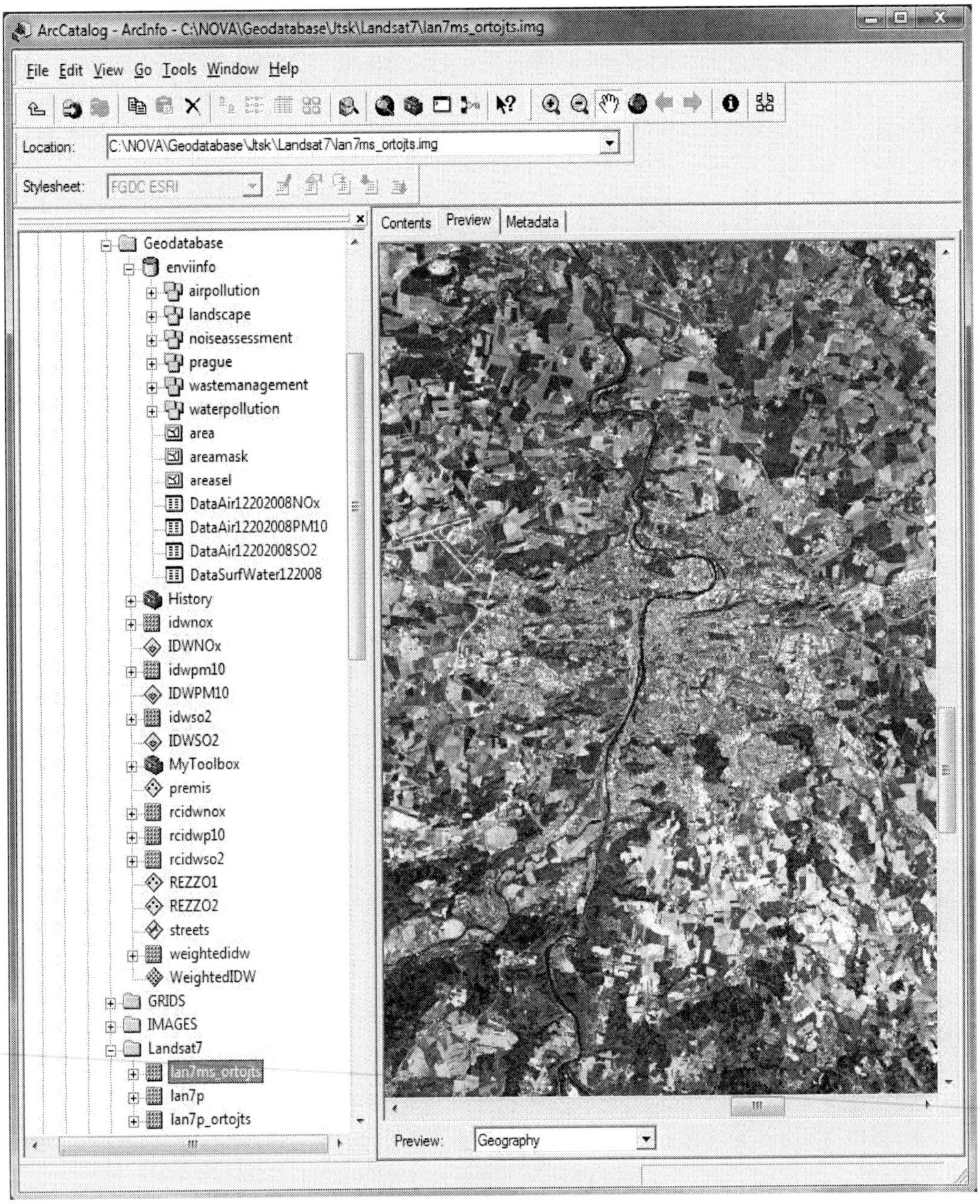

Figure 1. The data structure of the GIS project for estimating of the environmental effects: the list of the map layers is on the left site, the whole city from the Landsat 7 is in the preview mode.

As an example, measurements of air pollution are determined by monitoring of pollutants (SO_2, PM_{10}, NO_x) in the ground-level strata of atmosphere, and measurements of surface water quality are taken at fixed sampling points in the urban surface water network. Information about the ground-level air pollution is based on the public data accessible through the Internet (http://www.premis.cz , December, 2008). The actual configuration of the monitoring network is 15 monitoring stations that measure 12 main pollutants. Data are recorded in one hour period. The data are archived in the database of the Air Quality

Information System of the Czech Republic. The surface water quality assessment is performed according to the local standards. The recommended monitoring frequency is 12 samples taken per a year. Surface water is classified into five classes based on quality parameters over a longer period, usually a two-year period in order to have 24 data series (http://envis.prague-city.cz , December, 2008). The selected measurements are stored in the geodatabase (EnviInfo) and linked to the map layers. The network of sampling sites is in Figure 2, which also includes, as an example, measurements of the biological oxygen demand (BOD) in the attached column diagrams.

In case of air pollution, the prediction map layers for selected pollutants (SO_2, PM_{10}, NO_x) were created by examining the relationships between all of the sample points and producing a continuous surface of concentration. In order to produce the predictions maps, there are two main groupings of interpolation techniques: deterministic and geostatistical. In spite of that the geostatistical methods offer not only prediction surfaces but also error or uncertainty surfaces to give an indication of how good the predictions are, the deterministic method Inverse Distance Weighted (IDW) was selected considering to the limited number of the sample points-monitoring stations. This interpolation method forces the resulting surface to pass through the data value at sampling points and assumes that each measured point has a local influence that diminishes with distance. It weights the pollutant concentration at the points closer to the prediction location greater than those farther away. An example of the spatial interpolations based on IDW for PM_{10} concentrations is in Figure 3. Similarly, the spatial interpolation derived from the same sampling locations and based on IDW for NO_x is in Figure 4. In order to explore the results of the PM_{10} prediction map, the map composition in Figure 3 contains the map layers of the monitoring stations, the road network, and the potential sources of air pollution based on the national Air Pollution Sources Register (large and the mid-sized pollution sources: REZZO 1 and REZZO 2). The NO_x prediction map in Figure 4 is complemented by the map layers of the monitoring stations and the road network that is considered to have the high influence on the NO_x concentration in urban areas together with ozone and other compounds [31].

The next step in Figure 5 demonstrates using of the GIS spatial analysis to produce a suitability map by combining prediction maps of air pollutants (NO_x, PM_{10}, SO_2). The prediction maps are reclassified to a common scale, weighted according to a percentage influence and, finally, combined to produce a map displaying suitable locations for residential building. As an example, the datasets from the air pollution compartment have been selected to document the GIS ability for creating the spatial weighting matrixes for prediction of the final rating. Similarly, the map layers from other environmental compartments dealing with water pollution, waste management, noise assessment, and monitoring of environmental impacts on population health can be rated or complement existing results [32, 33]. Other GIS rating can include the calculation of the traveling time to the closest services and the origin-destination cost matrixes. Multimodal network datasets in the GIS enable to model multiple forms of transport across rail stations, subway stations or bus stops that form the linkages between several different forms of transportation. The temporal changes can be identified by comparison of a few spatial weighting matrixes linked to the selected time events.

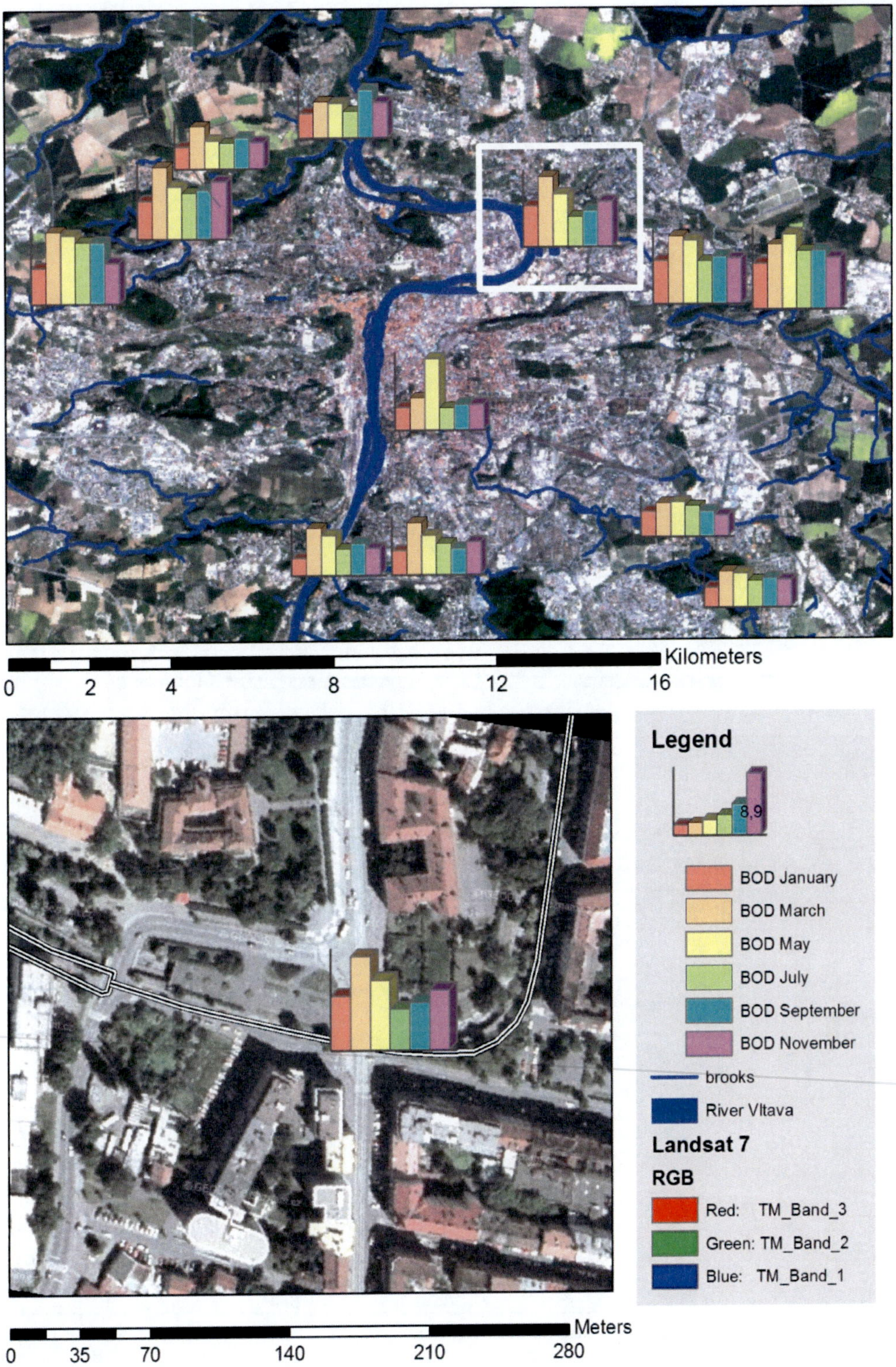

Figure 2. The measurements of water quality at the sampling points linked to the GIS layer.

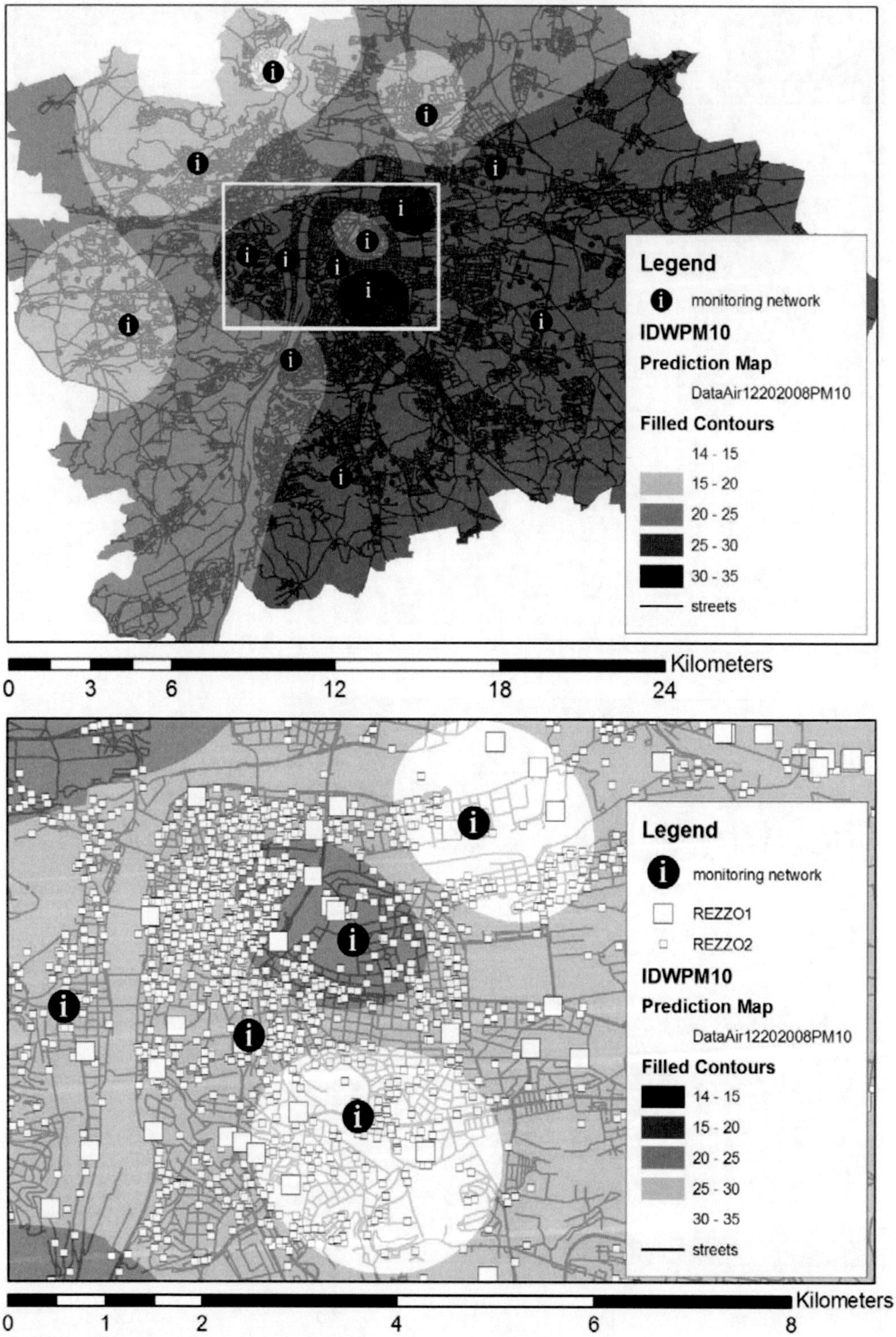

Figure 3. The prediction map of PM10 concentrations in the city Prague and its central part.

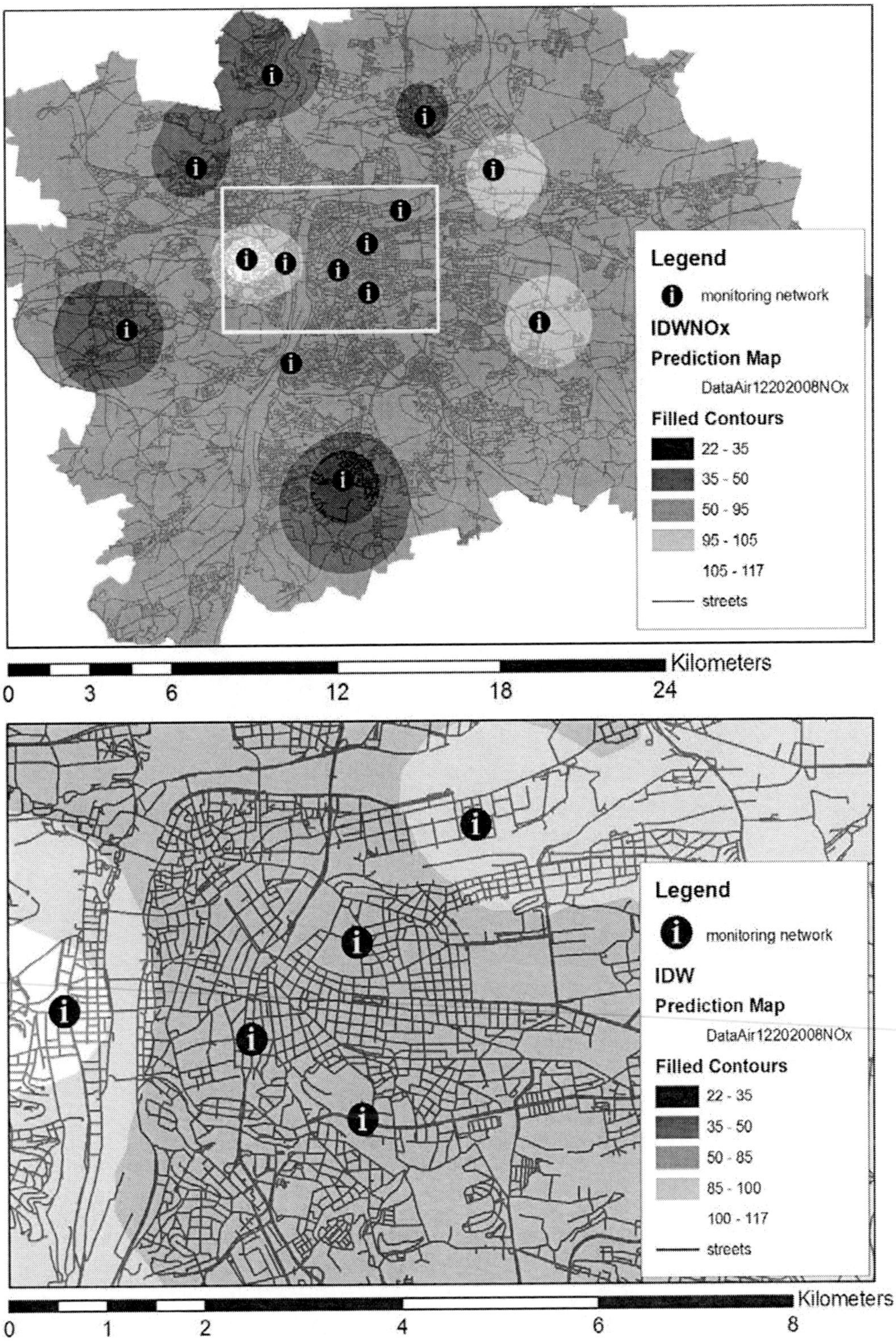

Figure 4. The prediction map of NOx concentrations in the city Prague and its central part.

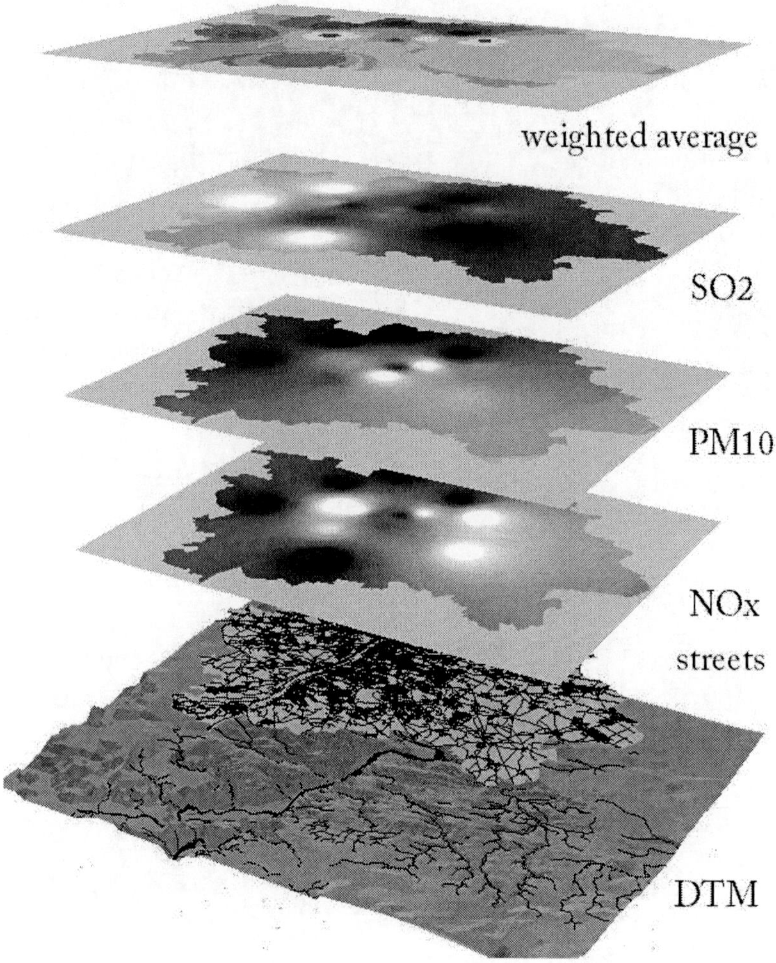

Figure 5. Mapping of suitable locations for residential building based on estimating the environmental effects: the light sites in the average weighted map layer indicate higher levels of mixed air pollution derived from spatial weighting matrixes for prediction of the final rating, the bottom layers represent the DTM with the draped surface water network and the street network.

6. Conclusion

The presented research shows various possibilities of using the GIS and its associated systems in the framework of estimating the environmental effects on residential property value. Considering to the complexity of this research, the attached case study is mainly focused on environmental compartments linked to the air pollution and partially to the surface water pollution. The spatial analysis demonstrates the weighting matrixes for prediction of the final rating, which can be extended to the whole range of environmental compartments and other phenomena. In order to provide the described complex analysis, a wide range of spatio-temporal data is needed to create a GIS project and to start database tasks. But, a great part of the digital data can be accessible from the U.S. EPA, landscape and urban planning departments, or local environmental agencies. Also, it is expected that the future GIS development will bring a better access to environmental data and insights to the research focused on building and environment.

ACKNOWLEDGMENT

The presented research was carried out in the framework of the project supported by Academy of Sciences in the Czech Republic (AVCR 1ET400760405). The spatio-temporal data and terrain measurements have been processed by ArcGIS, ERDAS Imagine in the GIS Laboratory at the Faculty of Natural Science, Charles University in Prague. The spatio-temporal data were captured via the Internet and from the annual reports with information on the Prague environment published by the city hall.

REFERENCES

[1] Cromley, E.K. and McLafferty S.L. (2002). *GIS and Public Health*. New York: The Guilford Press.

[2] Maguire, D.J., Batty, M., and Goodchild, M.F. (2005). *GIS, Spatial Analysis and Modeling*. Redlands, California: ESRI Press.

[3] Rigaux, P., Scholl, M. and Voisard, A. (2002). Spatial *Databases with Application to GIS*. San Francisco, California: Morgan Kaufmann Publishers.

[4] Mitchell, A. (2005). *The ESRI Guide to GIS Analysis. Volume 2: Spatial Measurements & Statistics*. Redlands, California: ESRI Press.

[5] Fedra, K. (1996). Distributed Models and Embedded GIS: Integration Strategies and Case Studies. In M.F. Goodchild, L.T. Steyaert, B.O. Parks (Ed.), *GIS and Environmental Modeling: Progress and Research Issues* (pp. 413-417). Fort Collins: GIS World, Inc.

[6] Tomlinson, R.F. (1987). Current and potential uses of geographical information systems: The North American experience. *International Journal of Geographic Information Systems, 1(3),* 203-218.

[7] Longley, P.A., Goodchild, M.F., Maguire, D.J. and Rhind, D.W. (2001). *Geographic Information Systems and Science*. New York: John Wiley & Sons Ltd.

[8] Johnston, C.A. (1998). Geographic Information Systems in Ecology. London: Blackwell Science Ltd.

[9] Zeiler, M. (1999). *Modeling Our World*. Redlands, California: ESRI Press.

[10] MacDonald, A. (2001). *Building a Geodatabase*. Redlands, California: ESRI Press.

[11] Zeiler, M. and Arctur, D. (2004). *Desiginig Geodatabases*. California: ESRI Press.

[12] Campbell, J.B. (2002). *Introduction to Remote Sensing*. London: Taylor&Francis.

[13] Li, Z., Zhu, Q. and Gold, C. (2005). *Digital Terrain Modeling: Principles and Methodology*. New York: CRC Press.

[14] Landgrebe, D.A. (2003). *Signal Theory Methods in Multispectral Remote Sensing*. New Jersey: John Wiley and Sons.

[15] Richards, J.A. and Jia, X. (2006). *Remote Sensing Digital Image Analysis*. Berlin: Springer.

[16] Stacey, S. and Pouncey, R. (1997). *Erdas Field Guide*. Atlanta: Erdas, Inc.

[17] Mower, B. (1998). A multiple source approach to acute human health risk assessment. *Waste Management 18*, 377-384.

[18] Bachman, W., Sarasua, W., Hallmark, S. and Guensler, R. (2000). Modeling regional mobile source emissions in a geographic information system framework. *Transportation Research Part C 8*, 205-229.

[19] El-Fadel, M., Shazbak, S., Baaj, M.H. and Saliby, E. (2002). Parametric sensitivity analysis of noise impact of multihighways in urban areas. *Environmental Impact Assessment Review 22*, 145-162.

[20] Yanosky, J.D., Paciorek, C.J., Schwartz, J., Laden, F., Puett, R. and Suh H.H. (2008) Spatio-temporal modeling of chronic PM_{10} exposure for the Nurses' Health Study. *Atmospheric Environment 42*, 4047-4062.

[21] Madsen, C., Carlsen, K.C., Hoek, G., Oftedal, B., Nafstad, P., Meliefste, K., Jacobsen, R., Nystad, W., Carlsen, K.H. and Brunekreef, B. (2007). Modeling the intra-urban variability of outdoor traffic pollution in Oslo, Norway-A GA^2LEN project. Atmospheric *Environment 41*, 7500-7511.

[22] Grivas, G., Chaloulakou, A. and Kassomenos, P. (2008). An overview of the PM10 pollution problem, in the Metropolitan Area of Athens, Greece. Assessment of controlling factors and potential impacts of long range transport. *Science of the Total Environment 389*, 165-177.

[23] Vardoulakis, S., Gonzalez-Flesca, N., Fisher, B.E.A. and Pericleous, K. (2002). Spatial variability of air pollution in the vicinity of a permanent monitoring station in central Paris. *Atmospheric Environment 39*, 2725-2736.

[24] Niemczynowicz, J. (1999). Urban hydrology and water management-present and future challenges. *Urban Water 1*, 1-14.

[25] Matejicek, L. (2006). Modeling of Water Pollution in River Basins with GIS. In A.R.Burk (Ed.), *Focus on Ecology Research* (pp. 247-270). New York: Nova Science Publishers.

[26] Leao, S., Bishop, I. and Evans, D. (2001). Assessing the demand of solid waste disposal in urban region by urban dynamics modelling in a GIS environment. *Resources, Conservation and Recycling 33*, 289-313.

[27] Azócar, G., Romero, H., Sanhueza, R., Vega, C., Aguayo, M. and Muñoz, M.D. (2007). Urbanization patterns and their impacts on social restructuring of urban space in Chilean mid-cities: The case of Los Angeles, Central Chile. *Land Use Policy 24*, 199-211.

[28] Koomen, E., Dekkers, J. and van Dijk T. (2008). Open-space preservation in the Netherlands: Planning, practice and prospects. *Land Use Policy 25*, 361-377.

[29] Culshaw, M.G., Nathanail, C.P., Leeks, G.J.L., Alker, S., Bridge, D., Duffy, T., Fowler, D., Packman, J.P., Swetnam, R., Wadsworth, R. and Wyatt, B. (2006). The role of web-based environmental information in urban planning-the environmental system for planners. *Science of the Total Environment 360*, 233-245.

[30] Pullar, D. and Tidey E. (2001). Coupling 3D visualisation to qualitative assessment of built environment designs. *Landscape and Urban Planning 55*, 29-40.

[31] Matejicek, L., Engst, P., Janour, Z. (2006). A GIS-based approach to spatio-temporal analysis of environmental pollution in urban areas: A case study of Prague's environment extended by LIDAR data. *Ecological Modelling 199*, 261-277.

[32] Matejicek, L., Janour, Z. and Strizik, M. (2008). Spatial Modeling of Air Pollution Based on Traffic Emissions in Urban Areas. In S.E. Paterson, L.K. Allan (Ed.), *Road Traffic: Safety, Modeling, and Impacts* (in print). New York: Nova Science Publishers.

[33] Matejicek, L. (2008). Spatial Modeling and Optimization of Municipal Solid Waste Collection in Urban Regions. In A.S. Martin, J.D. Suarez (Ed.), *Solid Waste: New Research* (in print). New York: Nova Science Publishers.

In: Environmental Modeling with GIS
Editor: Lubos Matejicek, pp. 129-144

ISBN: 978-1-60876-363-4

Chapter 9

SPATIO-TEMPORAL DATA MANAGEMENT FOR ENVIRONMENTAL MODELING OF DUST DISPERSION OVER OPENCAST COAL MINING AREAS

Lubos Matejicek

Institute for Environmental Studies, Charles University in Prague, Faculty of Natural Science, Benatska 2, 128 01, Prague, Czech Republic

ABSTRACT

There is a significant need for an integrated environmental information system containing both a knowledge base and decision-support tools to manage environmental protection of living environment near the opencast mining areas. Effective environmental management and research require spatio-temporal data to anticipate and predict the impact of the mining development. However, current data, while often extensive, are fragmented and inconsistent. Given that there already exists data repositories and collections of tools and modeling systems like continuous monitoring and dispersion modeling, the major contribution of the proposed GIS framework is to facilitate access to these items. An attached case study demonstrates GIS mapping of the opencast mining sites, and geodatabase management that supports spatial analysis and environmental modeling. A few spatial data inputs exist for building digital terrain models. The 3D laser scanner is used to acquire surface scans of the local sites. While the laser scans are used for mapping of the continuous parts, the differential GPS assists in the capturing of slope lines, surface points, and surface objects, such as mining facilities and residential buildings. In order to create the digital terrain model of the whole area of interest, all the input data are complemented by contour lines from existing digital map sources. For surface analysis and visualization of the opencast sites in a more realistic way, satellite images and aerial images are draped over the surface. This digital terrain model extended by data linked to emission sources, meteorological measurements and continuous air pollution monitoring represents the main input for spatial analysis and environmental modeling. Their outputs can form spatial interpolations, probability maps, wind flow layers, and isolines of pollutant concentrations based on dispersion modeling. All the data are managed by the ESRI's geodatabase. Each geodatabase data model represents both the ordered collections of simple features and rasters, as well as the rules and schema properties that add rich GIS behavior. It offers to use simultaneously several approaches to integrate spatio-temporally distributed environmental analyses and numerical simulations in the framework of GIS. They can range from simple linkage through shared data files to building environmental models as analytical functions into fully functional GIS. The geodatabase data models can also effectively be configured in a user-friendly environmental decision support system that is problem-specific and customized for environmental protection of residential zones.

Keywords: Spatio-temporal data; environmental modeling; GIS; opencast mining.

1. Introduction

Nowadays, there are several different contexts for the term modeling in the environmental sciences. Considering the scope of the traditional environmental modeling, the models are used to be representations of processes that are believed to occur in the real world, mainly from the hydrosphere, the geosphere, the atmosphere, and also from the biosphere. Thus, the vast variety of phenomena and approaches can be studied in the framework of environmental systems and their models. In consideration to the methods, environmental modeling can be deterministic or stochastic in dependence on nature of the environmental processes that are described by physical, chemical or biological laws and rules. They often include processes of diffusion, dispersion, advection, sorption and other basic phenomena that are applied to different environmental compartments, phases or scales. There are models that can simulate global processes of the entire globe or local phenomena of the mass and energy transport. The description of the processes and phenomena are based on the mathematical approach that often leads to the solution of differential equations representing the changes of variables, such as pollutant concentration or species density, in space and time. Now, the solution of such equations can be made by computer programs. Thus, a number of software packages can assist in time consuming numerical calculations of multidimensional problems, and spatio-temporal analysis of their results [1].

In the framework of spatial data processing by geographic information systems (GISs), a data model is considered as a template into which the data needed for a particular application can be fitted [2]. Thus, data models based on database design, mostly preferred for large data sets, can be also used for smaller data amounts. The GIS data models implemented in the environment of relational databases share universal concepts and best practices templates that are important for data sharing and building more complex spatio-temporal analyses [3]. For example, individual land cover classes can be represented by polygons, and lot lines demarcate their boundaries and share geometry with land cover classes according to a well-defined set of rules. Similarly, roads, railways or rivers are often represented as centerlines that are connected at their endpoints. The topological connection in a geometric network enables tracing and interconnection based on a set of rules. In addition to the geometric representation, all the features, such as polygons, lines and points, have a commonly used set of attributes and relationships to tax rolls. All data can be stored in the relation database that manages spatial features and supports spatio-temporal tasks.

But, in the framework of both the contexts, the models provide abstractions that simplify in communicating ideas to a range of stakeholders, from scientists to less scientifically informed public. The visual domain has become one of the most important ways of communicating spatio-temporal problems to a wide range of interests. The GISs with their visual forms in Euclidean space can simplify model representations to two- and three dimensional forms. Thus, GIS has become a powerful medium for communicating of spatio-temporal models in a wide variety of disciplines and from many different perspectives.

2. Spatio-Temporal Data Management

Actually, a number of new approaches in GISs and dynamic modeling bring challenges for integration of spatial data and time series data. The approaches are dealing with how objects and their aggregations interact and change in space and time. They identify different ways in which dynamics can be incorporated into models and suggest a series of numerical models that include environmental processes.

Defining of a spatial resolution is fundamental to the spatial representation, affecting not only the data storage but also the way of data visualization through analytical and simulation processes embedded in or coupled to the software. Spatial variance and correlation depend upon space determination, at what scale it is measured, and how and where the objects driving dynamics are located. With a GIS, the data can be modeled in three basic ways: as a collection of discrete features in the vector format, as a grid of cells in the raster format, or as a set of triangulated points included in the triangulated irregular network [4].

Modeling with vector data represented by features, such as points, lines, and polygons, is best applied to discrete objects with defined shapes and boundaries. The features are mostly digitized from map manuscripts or aerial photography, collected from GPS receivers, vectorized from raster data, or imported from other drawings. The points are stored as x, y coordinates in two dimensional layers, or as x, y, z coordinates in three dimensional layers. The lines are stored as paths of connected point coordinates. The polygons are stored as closed path. The points mostly represent small features or precise surface locations. The features with a length but a small width are formed by lines. Larger features that span an area are represented by polygons. In order to provide spatial analysis, line topology keeps track of which lines are connected to a node, and polygon topology keeps track of which polygons are their neighbors. Data models based on vector formats offer a wide range of spatial analyses, such as spatial and logical queries, buffer generation and proximity, polygon dissolve and overlay, or network analysis. Attributes of features can support analysis of temporal processes, such as variation of population density, or pollution concentration. Considering to the precise shape and position of the features, the vector data are appropriate for drawing.

Modeling with raster data represented by imaged or continuous data is mostly applied for continuous phenomena, such as elevation, pollution concentration, or harmful noise level. Each cell in the raster grid is a measured quantity. The raster dataset stores a two-dimensional matrix of cells with sampled values for each cell. Each cell has mostly the same width and height. Usually, the coordinates of the upper-left corner defines the location of the grid, and the cell size and the number of grid rows and columns determine the spatial extent of the raster dataset. The raster data often originate from aerial photography, satellite images, scanned photographs, or rasterized vector data. The raster data set can be stored as a matrix of numbers, which represent measured quantities. Thus, each cell is located by its row and column position. Small features or precise surface locations are represented by a single cell. The larger features in dependency on the cell size are formed by a series of adjacent cells with a common value. In case of the regular grids, the neighboring cells can be quickly located by their row and column values. Thus, raster data models are appropriate for spatial coincidence, surface analysis, spatial statistics and map algebra methods that can manage continuous features with gradually varying attributes.

The triangulated data representation is useful and efficient for capturing the surface of a piece of land. The data can be collected from GPS receivers, compiled from aerial photography, or converted from data that include elevation measurements. The triangulated irregular network (TIN) is stored as a set of nodes with x, y, z coordinates associated with their triangle and neighboring triangles. Z coordinates and the size of triangles determine the shape of a surface. Points define changes in the surface such as peaks. Ridges or streams can be defined by breaklines. Areas of exclusion, such as lakes, are determined by polygons with the same elevation. The TINs are best for estimation of elevations, slopes, aspects, volumes, vertical profiles, or viewshed analysis and rich presentation of surfaces.

Each of the described basic data models based on vector, raster, or triangulated network data formats is well suited for a particular class of environmental objects, and spatio-temporal analyses. These data models are not exclusive and can be integrated in the framework of advanced spatio-temporal data models. One of the tools for a data model design is represented by ESRI's geodatabase. The geodatabase schemas include definitions, integrity rules, and behavior for integrated datasets that include feature datasets, raster datasets, and so forth.

The feature dataset contains an organized collection of related feature classes, which enables to manage spatial relationships using topologies and networks. A feature class is represented by the same spatial elements, such as points, lines and polygons in vector formats. Topologies define how features share geometry, and control their integrity through the rules. Networks are used to model connectivity and flow between features.

Raster datasets can be organized into raster catalogs that bring a number of raster mechanisms to add behavior and to manage large multiuser raster collections. Considering to the evolution of database systems and GISs, the raster management can be actually provided in the framework of geodatabase.

In addition to feature datasets and raster datasets, the geodatabase can manage the triangulated data and a wide range of data structures based on formerly used data formats (shapefiles, coverages, or CAD data). Other user defined data structures can be included in order to support data processing based on information from monitoring systems or case oriented simulation tools.

One of the most important steps is designing a geodatabase schema before its creation [5]. Designing the schema is influenced by input data, spatial projections for data storage and display, rule definitions, future organization of object classes and their relationships, and many other requirements based on specific data inputs, selected spatio-temporal analyses and expected data outputs. A good design results in the well-constructed, functionally and operationally efficient database that satisfies objectives and organizational requirements. It represents the time consuming process that requires experience in GIS and database design.

3. Spatio-Temporal Data Sources

Spatio-temporal environmental data management can be generally considered as synthesis of spatial data and time series that originate from terrain observations, remote sensing, GPS mapping, monitoring systems, dynamic modeling and other disciplines focused on exploring of earth surface processes. In order to provide complex environmental analysis, GISs are often used as complex analytical tools for sharing spatio-temporal analytical methods and modeling tools.

3.1 Terrain observations

A number of datasets originate from terrain measurements provided by monitoring systems, or by collection of samples. Generally, the data collection can be divided to a few environmental compartments that are focused on air, water, landscape, waste, noise, and other phenomena. The air compartment contains emission sources classified into categories depending on the level of their effects on air quality. Additional data are represented by meteorological conditions, pollutant concentrations, and data from dispersion modeling. In order to carry out risk assessment, the data models need locations of emission sources, complex digital terrain models, basic meteorological data, and selected pollutant concentrations for model validations. The data series with emission levels and pollutant concentrations require spatio-temporal data management. Thus, the advanced data structures and rules are needed in the framework of a geodatabase design. Similarly, the water compartment can assist to model surface water pollution or groundwater pollution in dependence on pollution sources and other environmental parameters. The environmental compartment focused on landscape contains land cover data based on aerial photography, satellite images, and a wide range of landscape protection characteristics, such as areas of nature parks, areas of special protection, and registered significant landscape elements. The waste compartment can manage mapping of illegal dumpsites and registration of contaminated site in addition to municipal waste management. Nowadays, the exposure to excessive noise is one of the most serious factors affecting health not only inhabitants of larger cities, but also inhabitants of rural areas near the major roads, and the industrial sites. The long-term noise nuisance causes hearing impairments and a number of other diseases, such as stress and neurosis. Thus, the spatio-temporal analysis of long-term observation offers more complex information in addition to the noise map development.

3.2 Remote sensing

There are many concepts and definitions of remote sensing that evolve in dependence on implementation of new scientific discoveries and techniques. In general, remote sensing is the practice of deriving information about the Earth's land and water surfaces using images acquired from an overhead perspective, using electromagnetic radiation in one or more regions of the electromagnetic spectrum, reflected or emitted from the Earth's surface [6]. Information theoretic principles provide that information is potentially available at altitude from energy fields arising from the Earth's surface, and in particular from the spectral, spatial and temporal variations of those fields. The passive optical remote sensing systems may be divided into three basic parts: the scene, the sensor, and the processing system [7]. The scene lies in front of the sensor pupil, which consists of the Earth's surface and the intervening atmosphere. The sensor collects the main body of the data to be used. Besides widely used passive optical systems, although active microwave radar and optical lidar systems are important possibilities, in which case the illumination mechanism might be regarded as part of the sensor system. Considering to the spectrum of possible degrees of human/machine participation, the analysis process in the framework of the processing system can be divided into four parts: fully manual methods, machine aided manual methods, and automatic analysis. The methods are mostly focused on spatial data analysis and partitioning feature space. In case of environmental modeling of the dust dispersion, the supervised and unsupervised classifications are widely used to identify the potential area emission sources on the middle scale level, and to recognize individual objects, such as excavators, conveyors and temporally storage places, on the large scale level.

Relatively new methods in remote sensing are represented by implementation of lidar principles in the framework of surface terrain measurements. Nowadays, the 3D scanners provide an efficient method for the local 3D point cloud acquisition. In order to cover the whole area of interest with point cloud data, data acquisition is used to be provided from multiple standpoints. Then, the individual 3D point clouds are merged and rectified by remote sensing software or by advanced case oriented GIS tools. The output 3D point layers can significantly improve precision of the local sites in the digital terrain models (DTM) included in the GIS project.

Remote sensing images and terrain measurements represent important data sources for the GIS data models. The images are mostly included into raster datasets, the 3D point layers are often imported into vector datasets for development of the DTM. This short description indicates important data sources for development of data models for environmental modeling based on remote sensing, but the actual list of techniques in remote sensing is a quite more extensive.

3.3 GPS mapping

The precise GPS mapping is needed to locate surface terrain breaklines and peaks, man-made barriers, emission sources, and other surface objects. In order to improve the measurements, the data from the nearest reference stations are highly appreciated for accuracy improvement in the postprocessing phase. Thus, the original GPS measurements after the correction can be with the submeter accuracy.

Many recent GPS receivers are able to manage on-line corrections and can be used with a variety of field computers including laptops, Tablet PCs and PDAs. These computer tools are also able to run GIS programs that help to locate the areas of interest, display raster and vector layers, manage temporal data about explored objects, and provide basic spatial analysis. In addition to the GIS functionality, the developers can modify the user interface, optimize data sharing for updating of existing information in data models, and build macros for basic calculations and model simulations. Thus, the integration of the basic GIS functionality provides users with the ability to manage observations easily and efficiently [8].

3.4 Monitoring systems and spatio-temporal modeling

Data from monitoring systems are needed for development of dynamic models that are used to support decision-making processes. The systematic assessment of environmental characteristics, such as air quality or surface water pollution, based on regular measurements together with model predictions, can help to optimize living environment and improve public health. A set of tools for monitoring is created and upgraded in the territory of many countries. Considering to the huge amount of data, a number of assessment studies, research reports, and other specific documents based on legislative rules and requirements are published to inform research society, and public about environmental protection. In Europe, the national legislation of many countries is based on general approaches and sets of the limit values. The assessment is mostly carried out by measurements in areas where the level of air pollution, water pollution, etc. reaches the upper assessment thresholds, in dependence on the category of pollution, in short term or long term periods. Because the risk assessment must cover the whole area of interest, the modeling tools and analytical methods are needed for estimates of pollutant transport and accumulation in the surroundings of the monitoring stations or the sampling sites. The modeling research requires extensions of spatio-temporal data models, in order to manage model parameters, simulation outputs and input data for validation and spatial visualization.

Risk assessment focused on air pollution can be supported by a number of geostatistical tools, such as exploratory spatial data analysis (ESDA), semivariogram modeling, threshold mapping, model validation and diagnostics, or surface prediction using cokriging [9, 10, 11]. Dispersion modeling based on U.S. EPA software tools provide a more complex insight to the spatial and time scale of pollutant dispersion [12, 13, 14]. The other environmental compartments, such as surface water pollution, groundwater pollution, noise assessment, can be also explored by geostatistical tools, but there are many case oriented software systems in environmental modeling that assist in solution of particular tasks. The individual assessment tools include both the statistical methods and the dynamic models [15, 16, 17, 18, 19].

4. A Case Study: A Spatio-Temporal Data Model for Environmental Modeling of Dust Dispersion over Opencast Coal Mining Areas

In order to demonstrate spatio-temporal data management in the GIS, a case study focused on the ESRI's geodatabase data model was developed to support mapping of opencast mining sites, spatial analysis, and environmental modeling. Basic structure of the data model is in Figure 1. It includes data in the raster layers, such as aerial photographs, satellite images from Landsat 7, and aerial images from Google Earth. But, the largest part is formed by the vector data, such as thematic map layers, measurements by GPS, lidar scans in the 3D point clouds, and map layers with data from monitoring systems and dispersion modeling. The triangulated data are used for the DTM of the whole area of interest. The partial inputs for the data model are satellite images from Landsat 7 in Figure 2 and in Figure 3, and the 3D clouds of points captured by the laser scanner in Figure 4. As examples, the pollutant dispersion is displayed by GIS in Figure 5, and by Google Earth in Figures 6.

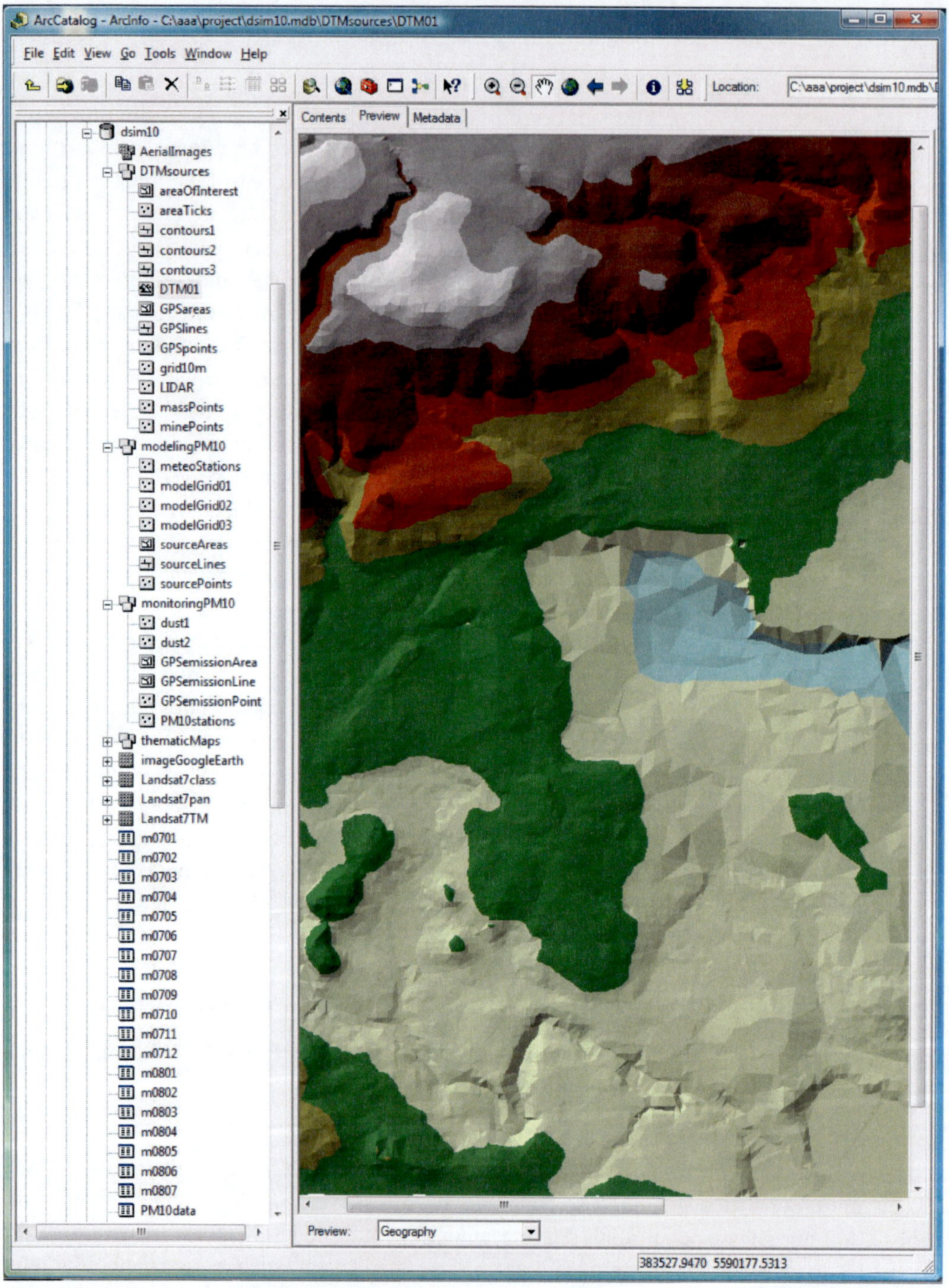

Figure 1. Basic structure of the data model with the raster, vector, and triangulated data for the opencast mining area. The list of the map layers and data tables is on the left side. It includes all the data needed for environmental modeling of dust dispersion over the opencast mining area. The short view on the DTM is on the right side. The data model is based on the ESRI's geodatabase, which offers complex integration of spatio-temporal data and rule definitions. In addition to the integration, data can be effectively managed and analyzed by a wide range of software tools included in the ESRI's ArcGIS-ArcCatalog and cooperating systems.

Figure 2. The satellite image from Landsat 7 TM (May 2000, bands 3-2-1, 30 metre pixels in the original image). The image also includes the location of monitoring stations and the surface water network for better identification of the areas of interest, the opencast mining sites. The opencast mining sites are represented by white color in the middle part of the image.

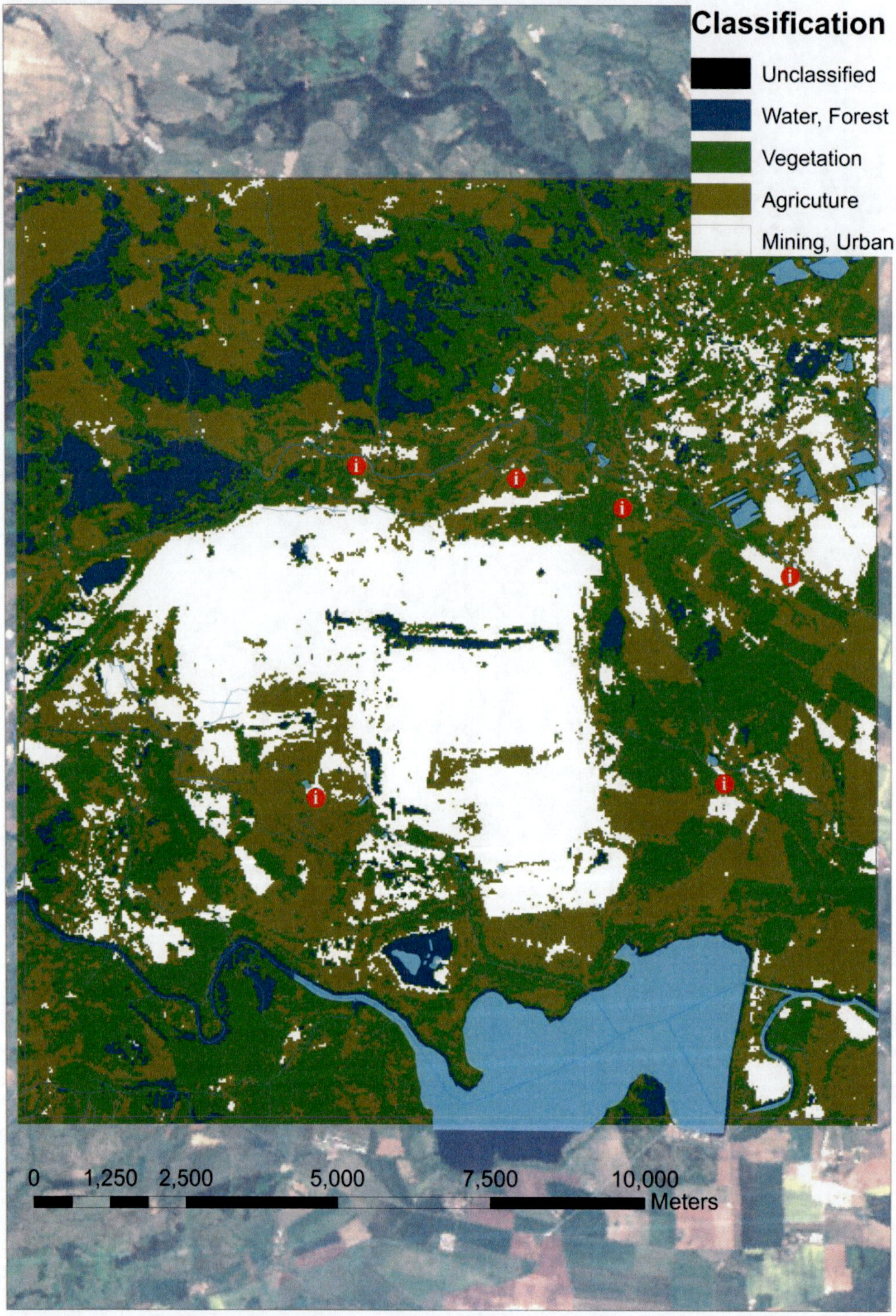

Figure 3. The classified satellite image from Landsat 7 TM. The image also includes the location of monitoring stations and the surface water network. The opencast mining areas and urban/industrial zones are indicated by white color.

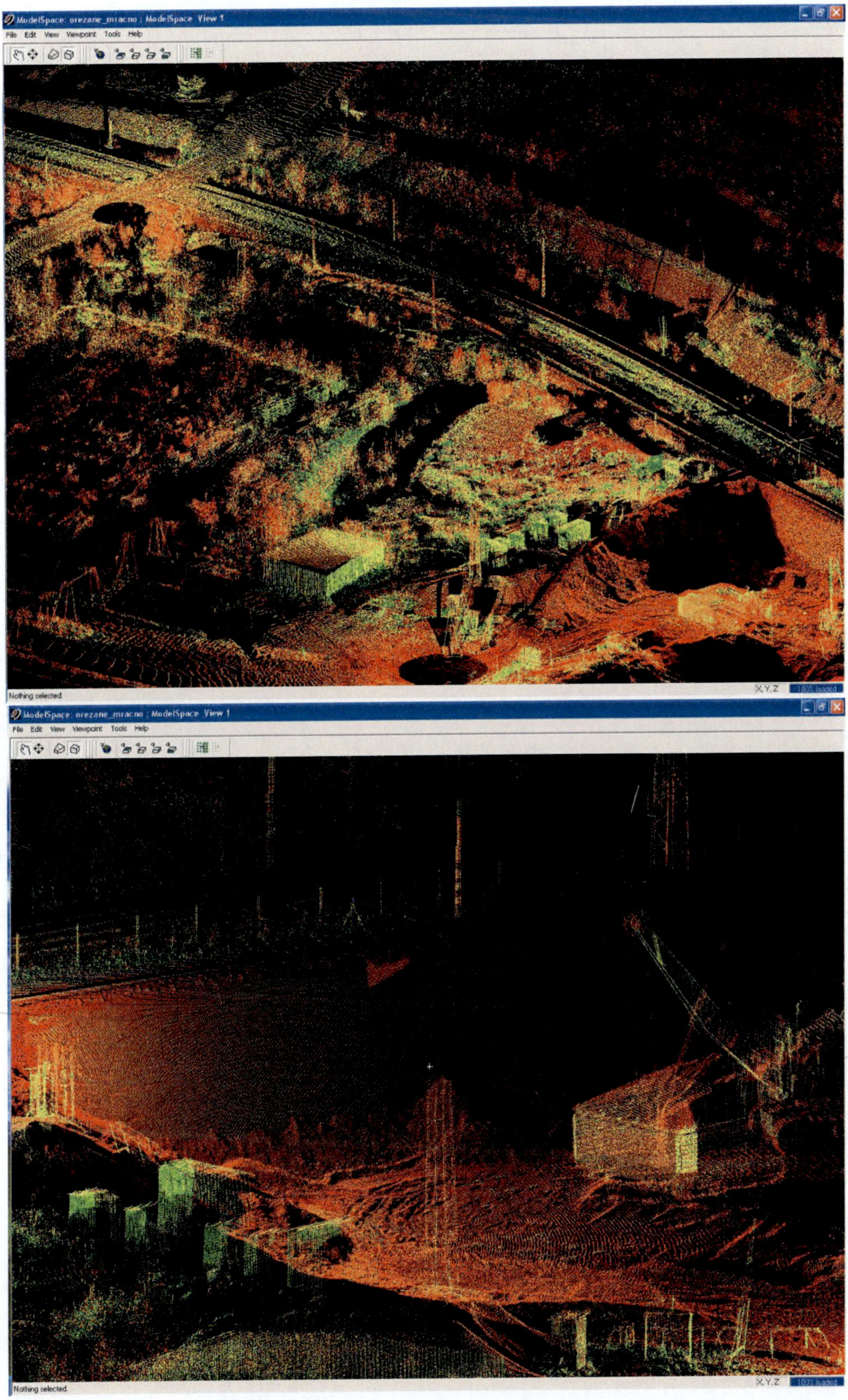

Figure 4. The merged and rectified 3D point clouds captured by the laser scanner Leica HDS 3000.

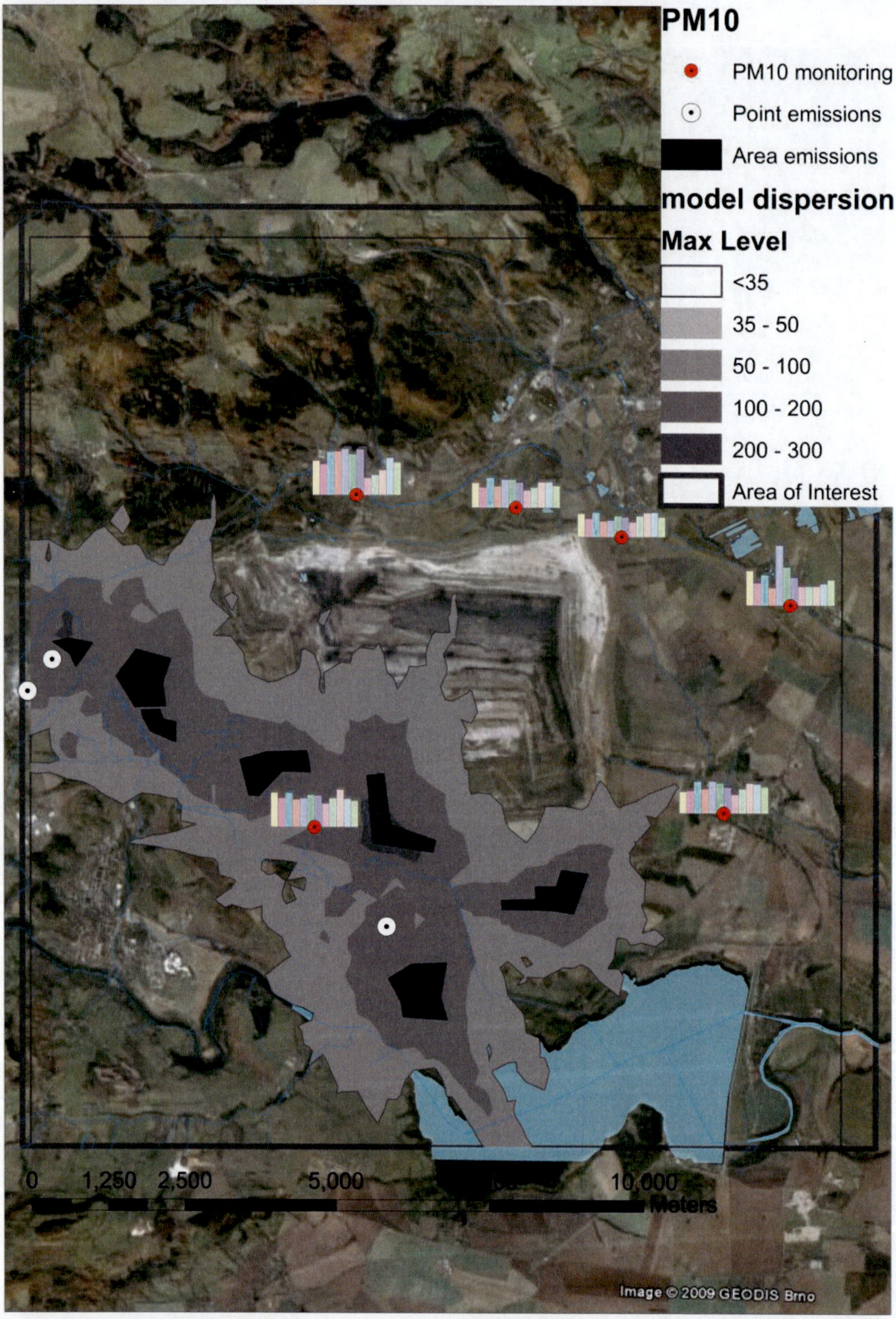

Figure 5. The filled line contours of the PM_{10} concentration based on the Cartesian receptor grid that represents the data output from the ISCST3 model. The contour map layer is complemented by the aerial photograph, by the layers with emission sources, and by the layer with charts of the average month PM_{10} concentrations (in micrograms, one year period) at the positions of the monitoring stations.

The area of interest with opencast mining sites based on the aerial images from Google Earth.

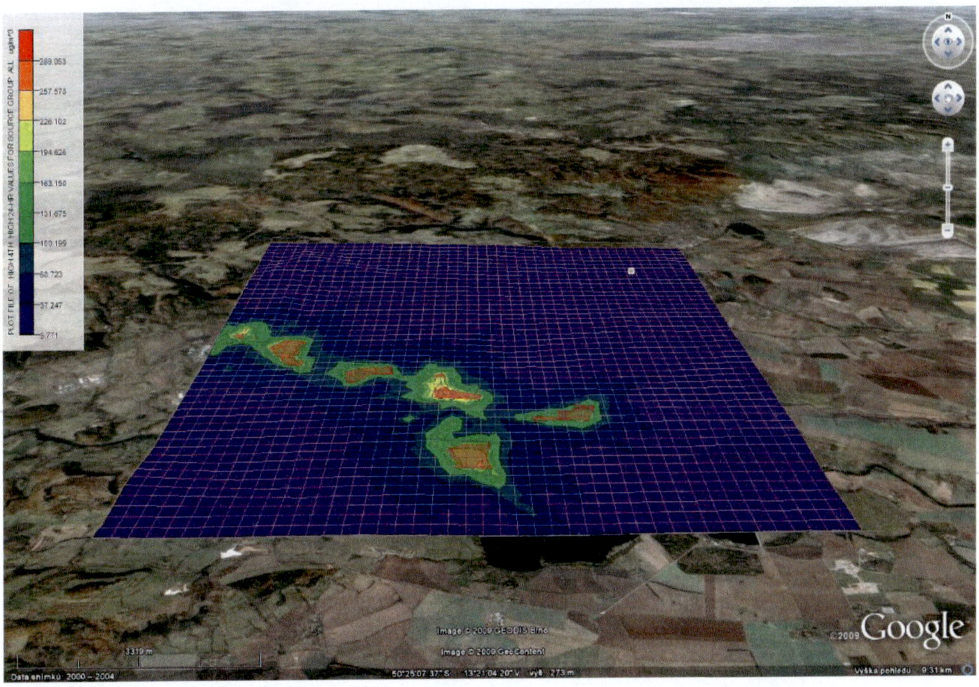

The area of interest complemented by the 3D projection of the aerial image from Google Earth, and by simulation outputs from the ISCST3 model. The simulation outputs include the receptor grid and the contour map layer. In order to visualize the simulation results, the data from modeling tools were exported from GIS and modeling tools and loaded into the viewer.

Figure 6. Visualization of the spatio-temporal data focused on dispersion modeling in Google Earth.

5. Discussion

The approach described in this contribution deals with spatio-temporal data models based on ESRI's geodatabase. In spite of that many sophisticated approaches contain spatio-temporal data models on various levels of complexity, using the spatial database tools in the framework of GIS enables more efficient data management extended by spatio-temporal analytical tools. Thus, a wide range of data formats can be shared for spatial analysis, such as data from terrain observations, thematic mapping, aerial photographs, satellite images, lidar scanning, GPS measurements, meteorological monitoring, and dynamic modeling. Then, the analytical tools can provide spatial analyses, such as ESDA, spatial interpolations based on deterministic and geostatistical methods, probability mapping, or cell based modeling. In addition to the spatial analysis, the temporal data can support dynamic models implemented in the standalone software tools, or GIS built-in modeling functions. It also documents recent research focused on spatio-temporal data, and dust dispersion modeling [20, 21].

As an example, the dispersion modeling of PM_{10} over the opencast mining sites by U.S. EPA software tools and GIS is used to demonstrate spatio-temporal data management, and sharing data together with standalone programming tools. In the framework of a GIS project, the spatio-temporal data model was developed to show new approaches in GISs and dynamic modeling. In addition to the standard thematic digital maps, the satellite image form Landsat 7 TM is used for classification of the surface areas, and identification of the potential erosion sites that can represent the area emission sources for dispersion modeling. In case of the presented dispersion modeling, the ISCST3 in the framework of the AERMOD View solves the PM_{10} concentration grid in the neighborhood of the area emission sources and the point emission sources. For this study, the dominant emission sources were finally located by GPS (coal power stations, and coal storage places). Determination of the emission rate was based on local predictions, and published empirical formulas [22]. But, corrections of the model parameters had to be provided in the framework of the model validation.

The presented various types of spatio-temporal data and case simulation data outputs in the GIS demonstrates the complexity of the proposed data model, which was developed in the ESRI's geodatabase environment. The attached case study also includes the complex DTM that is based on a wide range of spatial data managed by the spatio-temporal data model, which also documents its versatility.

6. Conclusion

Presented spatio-temporal data management with the spatio-temporal data model based on the geodatabase is made for environmental decision-making, which is extremely complex due to intricacy of the systems considered and the competing interests of the mining authorities, public, and research community [23]. Thus, additional development is usually needed to provide implementations and analytical extensions for an individual community. It involves development of integrated models, optimization algorithms and multicriteria decision analysis. Besides these case oriented tools, there is also a need to consider risk assessment in an integrated fashion for dealing with community and regional-based environmental problems.

The GIS and its growing influence on spatio-temporal data processing in environmental science guarantee the ongoing development of presented spatio-temporal data models and their future implementations in the framework of risk assessment.

ACKNOWLEDGMENT

The presented research was carried out in the framework of the project supported by Academy of Sciences in the Czech Republic (AVCR 1ET400760405). The spatio-temporal data and terrain measurements have been processed by ArcGIS, ERDAS Imagine, and AERMOD View in the GIS Laboratory at the Faculty of Natural Science, Charles University in Prague. The author is grateful to the following: the Manager, Emissions and Immisions Measurement Laboratory, Jan Brejcha and other members of the Brown Coal Research Institute in the Czech Republic for giving the sample data for local dispersion processes exploration, and help during the data processing.

REFERENCES

[1] Holzbecher, E. (2007). *Environmental Modeling Using MATLAB*. Berlin: Springer.

[2] Maguire, D.J., Batty, M. and Goodchild, M.F. (2005). *GIS, Spatial Analysis and Modeling*. Redlands, California: ESRI Press.

[3] Arctur, D. and Zeiler, M. (2004). *Designing Geodatabases: Case Studies in GIS Data Modeling*. Redlands, California: ESRI Press.

[4] Zeiler, M. (1999). *Modeling Our World*. Redlands, California: ESRI Press.

[5] MacDonald, A. (2001). *Building a Geodatabase*. Redlands, California: ESRI Press.

[6] Campbell, J.B. (2002). *Introduction to Remote Sensing*. 3rd edition. New York: Taylor & Francis.

[7] Landgrebe, D.A. (2003). *Signal Theory Methods in Multispectral Remote Sensing*. New Jersey: John Wiley and Sons.

[8] Matejicek, L. (2003). Development of Software Tools for Ecological Field Studies Using ArcPad. http://gis.esri.com/library/userconf/proc03/p0333.pdf

[9] Cressie, N.A.C. (1993). *Statistics for Spatial Data*. New York: Wiley.

[10] Johnston, K., VerHoef, J.M., Krivoruchko, K. and Lucas, N. (2001). *Using ArcGIS Geostatistical Analyst*. Redlands, California: ESRI Press.

[11] Matejicek, L., Engst, P. and Janour, Z. (2006). A GIS-based approach to spatio-temporal analysis of environmental pollution in urban areas: A case study of Prague's environment extended by LIDAR data. *Ecological Modelling 199*, 261-277.

[12] Wang, L., Parker, D.B., Parnell, C.B., Lacey, R.E. and Shaw, B.W. (2006). Comparison of CALPUFF and ISTST3 models for predicting downwind odor and source emission rates. *Atmospheric Environment 40*, 4663-4669.

[13] Puliafito, E., Guevara, M. and Puliafito, C. (2003). Characterization of urban air quality using GIS as a management system. *Environmental Pollution 122*, 105-117.

[14] Matejicek, L., Janour, Z., Benes, L., Bodnar, T. and Gulikova, E. (2008). Spatio-temporal modelling of dust transport over surface mining areas and neighbouring residential zones. *Sensors*, 3860-3847.

[15] Niemczynowicz, J. (1999). Urban hydrology and water management-present and future challenges. *Urban Water 1*, 1-14.

[16] Leao, S., Bishop, I. and Evans, D. (2001). Assessing the demand of solid waste disposal in urban region by urban dynamics modelling in a GIS environment. *Resources, Conservation and Recycling 33*, 289-313.

[17] Matejicek, L. (2006). Modeling of Water Pollution in River Basins with GIS. In A.R.Burk (Ed.), *Focus on Ecology Research* (pp. 247-270). New York: Nova Science Publishers.

[18] Matejicek, L. and Janour, Z. (2008). Spatio-Temporal Modeling of the Dust Emissions from an Opencast Coal Mining Area. In A.N. Dubois (Ed.), *Soil Contamination: New Research*. New York: Nova Science Publishers.

[19] Matejicek, L. (in print). Modeling of Traffic-Related Environmental Pollution in the GIS. In S. Demidov, J. Bonnet (Eds.), *Traffic Related Air Pollution and Internal Combustion Engines*. New York: Nova Science Publishers.

[20] Shaw, W.J., Allwine, K.J., Fritz, B.G., Rutz, F.C., Rishel, J.P. and Chapman, E.G. (2008). An evaluation of the wind erosion module in DUSTRAN. *Atmospheric Environment 42*, 1907-1921.

[21] Deutsch, F., Vankerkom, J., Janssen, L., Janssen, S., Bencs, L., Van Grieken, R., Fierens, F., Dumont, G. and Mensink, C. (2008). Modelling concentrations of airborne primary and secondary PM_{10} and $PM_{2.5}$ with the BelEUROS-model in Belgium. *Ecological Modelling 217*, 230-239.

[22] Chakraborty, M.K., Ahmad, M., Singh, R.S., Pal, D., Bandopadhyay, C. and Chaulya, S.K. (2002). Determination of the emission rate from various opencast mining operations. *Environmental Modelling & Software 17*, 467-480.

[23] Ascough II, J.C., Maier, H.R., Ravalico, J.K. and Strudley, M.W. (2008). Future research challenges for incorporation of uncertainty in environmental and ecological decision-making. *Ecological Modelling 219*, 383-399.

In: Environmental Modeling with GIS
Editor: Lubos Matejicek, pp. 145-159

ISBN: 978-1-60876-363-4

Chapter 10

SPATIO-TEMPORAL MODELING OF ENVIRONMENTAL STRESS IMPACTS ON VEGETATION CHANGES IN MOUNTAIN FOREST

Lubos Matejicek
Institute for Environmental Studies, Charles University in Prague, Faculty of Natural Science, Benatska 2, 128 01, Prague, Czech Republic
Eva Vavrova
Institute for Environmental Studies, Charles University in Prague, Faculty of Natural Science, Benatska 2, 128 01, Prague, Czech Republic

ABSTRACT

Innovative techniques, properly used, can offer exciting opportunities for the advancement of ecology. In case of geographic information systems (GISs), ecologists have a powerful and rapidly growing set of methods and tools with which to solve fundamental problems of a theoretical and applied nature. Previous analytical methods used in these subdisciplines have become implemented into GIS software. Current research in the field of ecology is increasingly integrated with the evolution of GPS, geostatistics, remote sensing and other techniques for data collection and spatial analysis, which promotes the use GIS. As an example, the data model for forest management is introduced and other extensions for spatio-temporal modeling are discussed. It is supported by the ESRI's geodatabase that can manage geographic data for GIS. The described forestry data model originates from research results created by forestry users attending the 1999 ESRI Users Conference in the framework of presentations of GIS and the geodatabase technology. The areas of interest for implementation of the data model include the White Mountain National Forest and the mountain Norway spruce forests that belong to the most endangered Central-European forest ecosystems.

Keywords: Spatio-temporal modeling; forestry; GIS.

1. INTRODUCTION

A new computer technology creates the need for innovation of a wide range of methods and approaches that are available to ecologists. Innovative techniques, properly used, can offer exciting opportunities for the advancement of ecology. In case of geographic information systems (GISs), ecologists have a powerful and rapidly growing set of methods and tools with which to solve fundamental problems of a theoretical and applied nature. This text focuses on selected aspects of GIS and related technology that partially deal with modeling of floor vegetation changes in declining mountain Norway spruce forests. But, some described GIS concepts and methods are not specific to any discipline.

In spite of that GISs are often considered the domain of geographers, a myth that tends to be perpetuated by geographers and ecologists alike. One of the GIS advantages is the ability to analyze questions of a spatial nature that can explore the location of a biological entity relative to other organisms or the environment influences its functioning. Traditionally, the GIS is used to explore features on the Earth's surface, but its spatial data management and analytical capabilities can be used with any spatially distributed data. For example, GIS can assist in spatial distribution of floor vegetation in the local scales, tree inventories for spatio-temporal modeling, insect damage to leaves, etc. Thus, the GISs can process data at a variety of spatial scales, from global to microscopic. The GIS is also related to other software tools, such as databases, statistical programs, image processing, and computer aided design (CAD). The recent GISs are mostly linked with the database that is managed by the Relation Database Management System (RDBMS). While the RDBMS organizes and query data, GIS contains functions for spatial analysis and display that can manage data in the database by Structured Query Language (SQL). Likewise the database is used for management and sharing data, statistical packages are used for more advanced analysis, because the GIS is mostly focused only on rudimentary statistical operations. Other data sources for GIS represent digital images from remote sensing and surface locations from Global Positioning System (GPS). Satellite images acquired as a matrix of pixels have growing importance for environmental studies. The most advanced imagery products, such as data from GeoEye-1, have a resolution of 0.41 meters (0.50-meters for commercial customers). This sub-meter ground resolution imagery discerns an object on the ground approximately 16 inches in size using both multispectral imagery and panchromatic imagery. In case of GPS, the recent advanced receivers for GIS can achieve 10 cm (4 inch) accuracy by combining various new principles and technology (H-Star™ technology, OmniSTAR support, and with the option of GLONASS support on top of dual frequency GPS).

In the framework of ecological research, many GIS applications have been carried out for a few last decades. Officially, GIS began to be adopted by U.S. ecologists in the late 1980s by featuring in a special symposium held at the 1987 by Ecological Society of America Annual Meeting in Davis, California, which continued in subsequent symposiums and conferences [1]. In addition to publishing papers presented in this symposium, a tremendous volume of scientific research outputs based on GIS is annually presented in scientific papers, international conferences, and reports. GIS has become a standard tool in landscape ecology. Many techniques originated from ecological research have become incorporated into the GIS software tools. Nowadays, GIS also becomes to be implemented in the field computers that can manage GPS information and data from external sensors.

2. GIS for Quantitative Theory and Modeling

Modeling with GIS is transforming real objects and processes into the way we describe and study the earth. In addition to spatial data storage and query, the GIS has to integrate a wide range of data, because the environment is connected to natural, social, and economic processes that operate on local, regional, and global scales. Likewise in environmental modeling, contaminants and biological phenomena are not usually concentrated at a single point. In these cases, GISs have proven to be powerful tools for modeling environmental conditions across the full ranges of geographical scales, from local to global. Many GIS projects model temporal and spatial patterns of hazards for environmental health problems, which are important for public health intervention activities [2].

A number of conceptual, analytical, and technological advances in the last century have contributed to the development of GIS for environmental applications. For example, Harvard's laboratory for computer graphics was one of the leading workplaces for GIS in the early 1960s. The SYMAP computer program was the prototype digital mapping system designed for spatial data analysis. The program was adapted to process point and area data, interpolate a surface, and create matrixes. Two subsequent computer programs, GRID and IMGRID, were focused on raster data. Vector data objects were consequently processed in the computer program ODESSEY. Furthermore, several major land use inventories have been developed since the 1970s. The first national scale GIS was the Canada GIS (CGIS), which was implemented in 1964 to analyze Canadian Land Inventory data. This system was followed by other GISs, mostly conducted for natural resources inventory in the U.S.A. An early vector based GIS developed by the state of Minnesota, PIOS, served as the foundation for ARC/INFO software, which is actually implemented in the ArcGIS software developed by Environmental Systems Research Institute (ESRI). Other software applications in the area of GIS are represented by IDRISI, which was initially developed for teaching GIS courses at Clark University (Clark Labs for Cartographic Technology and Geographic Analysis) in the U.S.A. IDRISI is still widely used for GIS training and applications. Subsequently, national institutions and commercial firms started the development of GIS in a wide range of disciplines. In addition to the needs of digital maps at national scales, other progress in remote sensing and global position systems requires to reflect new functionality in GIS.

Major advances in GIS development were often enabled by computer hardware advances. Disk storage was one of the first major technological steps to benefit GIS development. Through the last century, the computers have decreased in physical size while increasing in power. So actually, GISs are programmed for smaller computer platforms, meaning minicomputers and especially microcomputers-PCs. PC workstations, with high processing power and large storage capacity, are becoming increasingly popular as GIS platforms. Their network connections can share data, allowing multiple users access to GIS projects to perform several operations simultaneously. GIS has been adopted by ecologists and environmentalists already for a few decades.

Previous analytical methods used in these subdisciplines have become implemented into GIS software. Current research in the field of ecology is increasingly integrated with the evolution of GPS, geostatistics, remote sensing and other techniques for data collection and spatial analysis, which promotes the use GIS [3, 4, 5]. Many new functions become to be implemented in the next versions of GIS, such as ESRI's ArcGIS.

3. The Need for Spatio-Temporal Forestry Models

As mentioned in the book about forest dynamics focused on ecological models written by D.B. Botkin [6], every science involves theory, and every theory involves models of how to object of study works. The models are usually used in two ways. The first way represents the conceptual models that are dedicated to explain how something works. The second way deals with formal models that are explicit in mathematical terms. Sometimes, ecological scientists apply mathematical models that are derived from other projects. In these cases, the assumptions and limitations of those models must be taken into account in order to provide proper conclusions.

Every model represents an intentional simplification of reality in order to examine, analyze and understand phenomena of interest. During of the first half of the twentieth century, there were dominant models of population growth, such as exponential and logistic models of the growth of a single population, and Lotka-Volterra models for predator-prey interactions [7, 8, 9, 10]. Other models tried to describe these phenomena by the physical principles of energy accumulation and transport [11, 12, 13, 14]. The details of many mathematical models have been reviewed and analyzed in many books and articles. But, in order to simplify the mathematical description, these models have serious limitations. They ignore the complex influence of the environment and of time-varying stochastic events.

It is clear that simple, continuous, differential equation models based on basic interactions and physical laws are no longer viable in forest ecology. Thus, new approaches have been carried out in the last decades. Some models were made possible by computer tools for simulation. In recent years there has been considerable growth of new technology, such GIS, GPS and new sensors in the framework of remote sensing [15, 16, 17, 18, 19].

In spite of that some models of tree growth using differential equations were developed in the nineteenth century, their numerical solutions and spatio-temporal data management were rather limited up to now. Nowadays, many models need to be revised in order to implement new research in new sensor technology, complex monitoring systems, and computer science. Generally, there is no one model for all purposes. Since the last century, the formal models of vegetation have been followed several lines of investigation : simple global models; systems analysis models; static stand structure models; simple population dynamics models; biophysical models; and physiological models [6, 20, 21, 22, 23, 24, 25].

Plant ecology had little systems theory and formal mathematical approaches during the first half of the twentieth century. The study of plant communities was rich in concepts about the dynamic patterns of change in vegetation over space and time. A strong emphasis in the framework of qualitative theory was on vegetation changes in climate with elevation or changes in soil conditions that accompanied differences in bedrock. This research provided an important background for the complex mathematical models, such as JABOWA [6], ZELIG [26], FORMIX [27], TREEDYN3 [28] and others [29, 30, 31, 32]. From a technical point of view, many models were initially implemented on mainframe computers, and lacked an integrated graphic display. Nowadays, integration of spatial and temporal data requires new computer systems based on the computer graphics and the network connection.

As an example, the area of the White Mountain National Forest, Figure 1-2, with Mount Washington, the highest peak in the northeastern U.S. is used for spatio-temporal modeling. Figure 3-4 illustrate the spatial data based on the digital terrain model (DTM).

Figure 1. The area of the White Mountain National Forest (Matejicek, June-2006).

Figure 2. Vegetation influenced by climatic synergistic stress impacts (Matejicek, June-2006).

Figure 3. Visualization of the area of interest by a digital terrain model with the draped image map over it. The scene is used for a 3D view of the area together with other map layers based on spatio-temporal modeling and simulation outputs from external case oriented modeling tools.

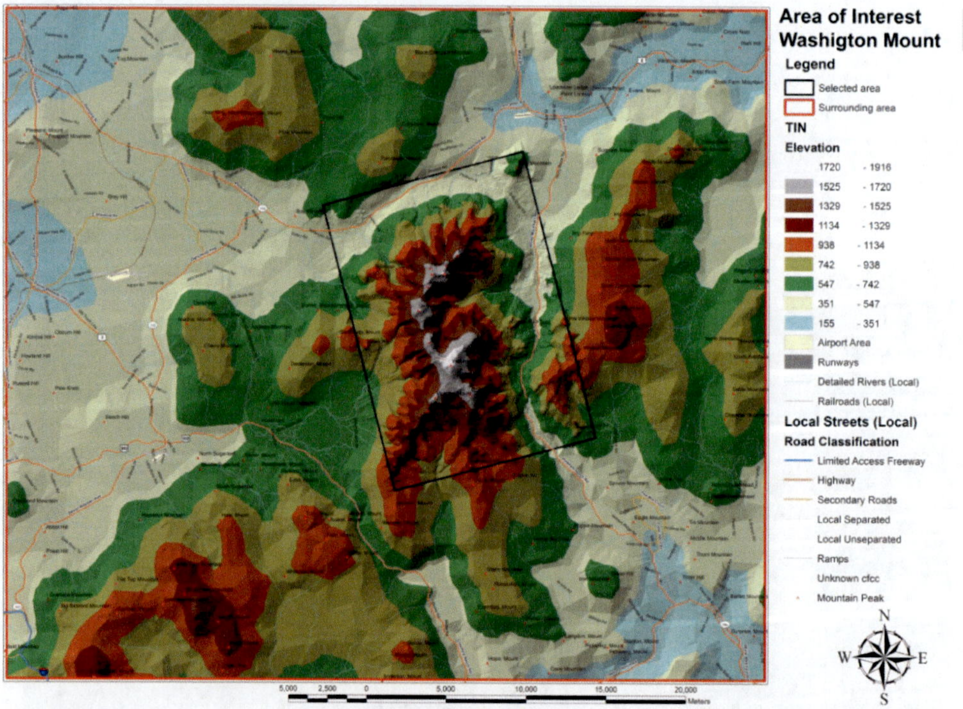

Figure 4. Digital terrain model of the area of the White Mountain National Forest with Mount Washington, the highest peak in the northeastern U.S., which is used for spatio-temporal modeling. The displayed map layers are managed by the geodatabase together with other data from monitoring systems, terrain measurements and case oriented simulation tools.

4. FORESTRY DATA MODELS

As an example, the data model for forest management is illustrated in Figure 5. It is supported by the ESRI's geodatabase that can manage geographic data for GIS [15, 16]. The presented forestry data model was created by forestry users attending the 1999 ESRI Users Conference in the framework of presentations of GIS and the geodatabase technology. New versions of the ESRI's ArcGIS have introduced an object-oriented technology to support a new data models for geographic information that can also include data structures for spatio-temporal modeling. The diagrams describing the data model are referred to as application design patterns, using the framework provided by the ESRI's ArcObjects. In addition to the benefit of communication, the data model diagrams created by case oriented tools, such as MS Visio or others, also provide the mechanism to create a database schema in the RDBMS. The Unified Modeling Language (UML) is an industry standard for supporting of this process. The objectives of the Forestry Spatial Interest Group (FSIG) to create the data model were:

-design an example geodatabase for forest management that can be modified in order to fulfill all the user requirements not only in forest management, but also in forestry spatio-temporal modeling,

-document the example in a data model diagram as an application design pattern that can stimulate other forestry approach in the framework of geodatabase design,

-implement a physical representation of the model as a prototype created within a personal geodatabase,

-populate the geodatabase with data from an operational system,

-produce supporting documentation,

-encourage FSIG members to critique the prototype and identify needed modifications.

It is assumed that the GIS implementation will vary due to differences in organizational context. Thus, the functions and decisions depend upon the management purpose of the organization, such as a government agency, a research group or a corporation (privately held or publicly owned). For example, the functions for a forest product company should include a production chain, with management constraints of space, time and resources. Annual harvest and production levels are constrained by limiting factors within each of the resources functions. Within this context, managers can address various types of decisions including operations, investment, and land stewardship. Operational planning for log production requires information on the appropriate equipment for the site, which is contributed by slope, soil conditions, and stand structure. Site productivity influences future growth of the stand and growth has a significant influence on management plans. Physical site characteristics are one source to estimate forest productivity. Management guidelines associated with the derived land classification combining multiple types of geographic data (solar exposure, soil nutrients, soil moisture and potential natural vegetation) can also match a treatment regime with the site and landscape conditions. Many other conditions and details are presented on the ESRI's website focused on downloading, creating and designing data model templates (http://support.esri.com). In case of the environmental stress impacts on vegetation changes in mountain forests, the similar rules can be applied in order to explore influence of limiting factors, potential air pollutants, and regeneration processes. For spatio-temporal modeling, other data structures are needed for sharing inputs/outputs of numerical models.

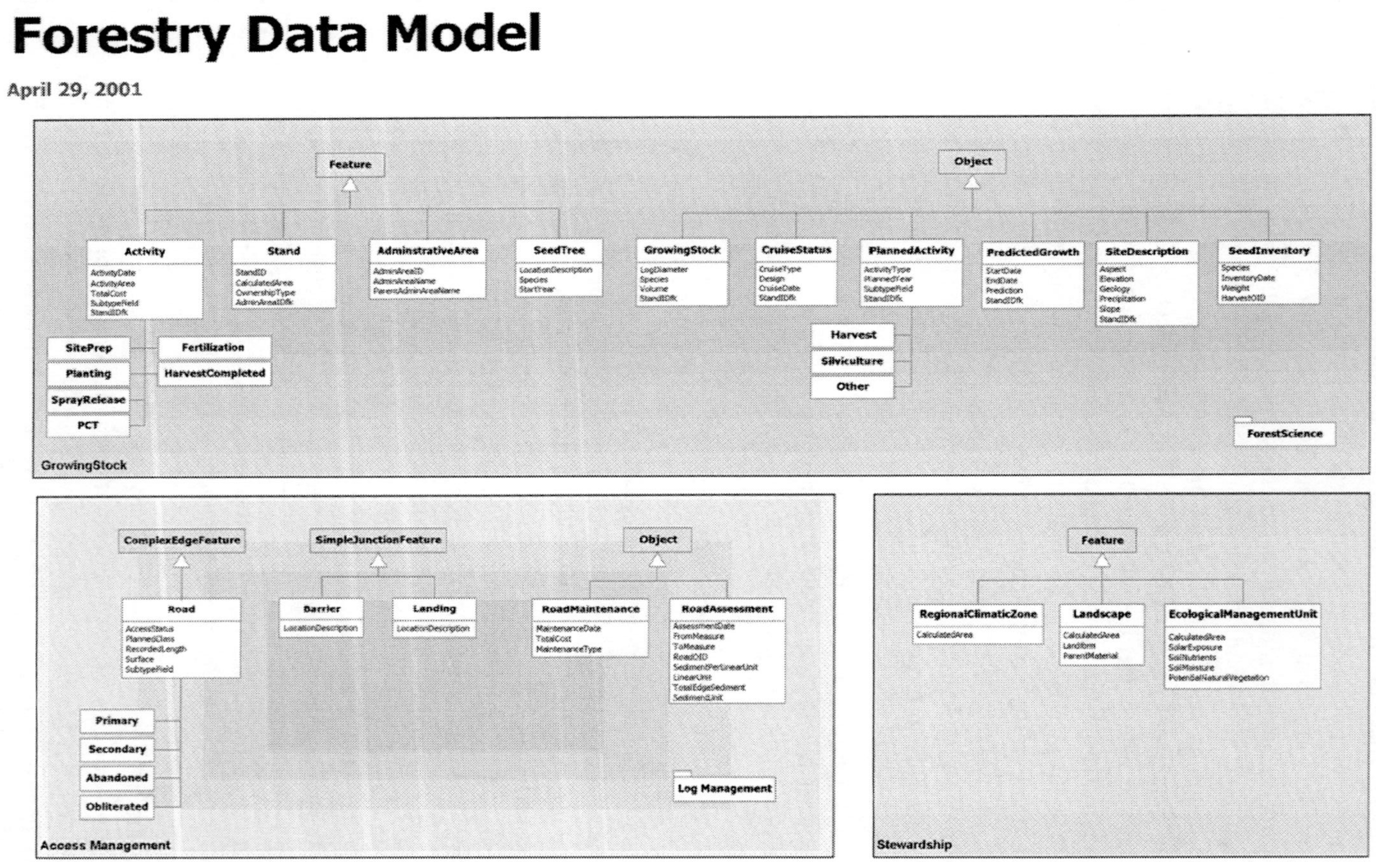

Figure 5. An example of a forestry data model for implementation in the ESRI's geodatabase.

5. A Case Study: Using GIS to Study Succession Dynamics of Floor Vegetation in Declining Mountain Norway Spruce Forest

Sufficient knowledge on natural regeneration is an important factor for maintaining the ecological stability of the mountain climax Norway spruce forests, which belong to the most endangered Central-European forest ecosystems. The synergistic effect of climatic and pollution stress impacts are considered to be a primary cause of a forest decline in this region. The decline of a tree layer resulted in increasing light supply into the stand and subsequently in expansion of the competitive floor vegetation species [33]. The aim of the presented study is to explore succession dynamics of floor vegetation during the decline of the mountain Norway spruce forests, and to process obtained data using the GIS [34, 35].

5.1 Materials and Methods

The research took place on five permanent research plots (50 x 50 meters) situated in autochthonous Norway spruce (*Picea abies*) stands in the Krkonoše Mts. in the period 1995-2009. These research plots differed mainly in the altitude (1185-1317 m), exposition, slope, soil type, pollution stress level and stage of the forest decline [36].

The GIS was used to carry out maps of particular soil coverage types, such as litter, rotten wood, mosses, *Avenella flexuosa*, *Calamagrostis villosa*, *Vaccinium myrtillus*, and other species. Spatial data objects were stored according to the soil coverage type into five layers: *Avenella flexuosa*, *Calamagrostis villosa*, *Vaccinium myrtillus*, other vegetation and a mixed layer with mosses, litter, rotten wood and other substrate.

Determination of spatial changes of the observed categories between 1995 and 2009 was processed by GIS's spatio-temporal methods. The significance of recognized changes in the period 1995-2009 was examined by statistical testing.

5.2 Results

Finally, a spatio-temporal model has been developed in order to predict changes of the observed categories. Thus, the partial differential equations were used to solve surface spatial interactions in the *x-y* directions, and competitions among species. Figure 6 contains fundamental growth equations for ground vegetation changes. All the variables and parameters are dependent on the time t and the location x, y. Thus, each local site has its specific setting of growth ratios and competition conditions for explored species in dependency on actual biotic and abiotic factors. All data are managed by the GIS, which offers more efficient and complex spatio-temporal modeling tools. As an example, Figure 7 shows one of the permanent plots that are focused on digital terrain models and tree layers (*Picea abies*). Figure 8 illustrates a map layer of tree crown projections and a map layer with selected local vegetation sites (1 x 1 meter). The GIS outputs, such as the map layers based on slope ratio and aspect ratio, are in Figure 9 and Figure 10.

Fundamental growth equations for ground vegetation changes in time (t) and space (x, y)

Dominant ground vegetation species *Avenella flexuosa* $\mathbf{s_{AF}}$:

$$\frac{\partial s_{AF}}{\partial t} = D_{AF}\frac{\partial^2 s_{AF}}{\partial x^2} + D_{AF}\frac{\partial^2 s_{AF}}{\partial y^2} + r_{AF}s_{AF}\left(1 - \frac{s_{AF} + \alpha_{CV}s_{CV} + \alpha_{VM}s_{VM} + \alpha_{GV}s_{GV}}{K_{AF}}\right)$$

Dominant ground vegetation species *Calamagrostis villosa* $\mathbf{s_{CV}}$:

$$\frac{\partial s_{CV}}{\partial t} = D_{CV}\frac{\partial^2 s_{CV}}{\partial x^2} + D_{CV}\frac{\partial^2 s_{CV}}{\partial y^2} + r_{CV}s_{CV}\left(1 - \frac{s_{CV} + \beta_{AF}s_{AF} + \beta_{VM}s_{VM} + \beta_{GV}s_{GV}}{K_{CV}}\right)$$

Dominant ground vegetation species *Vaccinium myrtillus* $\mathbf{s_{VM}}$:

$$\frac{\partial s_{VM}}{\partial t} = D_{VM}\frac{\partial^2 s_{VM}}{\partial x^2} + D_{VM}\frac{\partial^2 s_{VM}}{\partial y^2} + r_{VM}s_{VM}\left(1 - \frac{s_{VM} + \gamma_{AF}s_{AF} + \gamma_{CV}s_{CV} + \gamma_{GV}s_{GV}}{K_{VM}}\right)$$

Other ground vegetation $\mathbf{s_{GV}}$:

$$\frac{\partial s_{GV}}{\partial t} = D_{GV}\frac{\partial^2 s_{GV}}{\partial x^2} + D_{GV}\frac{\partial^2 s_{GV}}{\partial y^2} + r_{GV}s_{GV}\left(1 - \frac{s_{GV} + \delta_{AF}s_{AF} + \delta_{CV}s_{CV} + \delta_{VM}s_{VM}}{K_{GV}}\right)$$

Variables and parameters for ground vegetation changes in time (t) and space (x, y)

$\mathbf{s_{AF}}$	[m^2]	population size: *Avenella flexuosa*
$\mathbf{s_{CV}}$	[m^2]	population size: *Calamagrostis villosa*
$\mathbf{s_{VM}}$	[m^2]	population size: *Vaccinium myrtillus*
$\mathbf{s_{GV}}$	[m^2]	population size: other ground vegetation
D_{AF}	[$m^2 year^{-1}$]	coefficient of spatial propagation: *Avenella flexuosa*
D_{CV}	[$m^2 year^{-1}$]	coefficient of spatial propagation: *Calamagrostis villosa*
D_{VM}	[$m^2 year^{-1}$]	coefficient of spatial propagation: *Vaccinium myrtillus*
D_{GV}	[$m^2 year^{-1}$]	coefficient of spatial propagation: other ground vegetation
r_{AF}	[$year^{-1}$]	growth ratio: *Avenella flexuosa*
r_{CV}	[$year^{-1}$]	growth ratio: *Calamagrostis villosa*
r_{VM}	[$year^{-1}$]	growth ratio: *Vaccinium myrtillus*
r_{GV}	[$year^{-1}$]	growth ratio: other ground vegetation
K_{AF}	[m^2]	carrying capacity: *Avenella flexuosa*
K_{CV}	[m^2]	carrying capacity: *Calamagrostis villosa*
K_{VM}	[m^2]	carrying capacity: *Vaccinium myrtillus*
K_{GV}	[m^2]	carrying capacity: other ground vegetation
α	[-]	coefficients of competition: *Avenella flexuosa*
β	[-]	coefficients of competition: *Calamagrostis villosa*
γ	[-]	coefficients of competition: *Vaccinium myrtillus*
δ	[-]	coefficients of competition: other ground vegetation

Figure 6. Fundamental growth equations for ground vegetation changes, and a list of model variables and model parameters.

Figure 7. An example of the permanent plot (50 x 50 meters) with a digital terrain model and a tree map layer (*Picea abies*) that are situated in autochthonous Norway spruce.

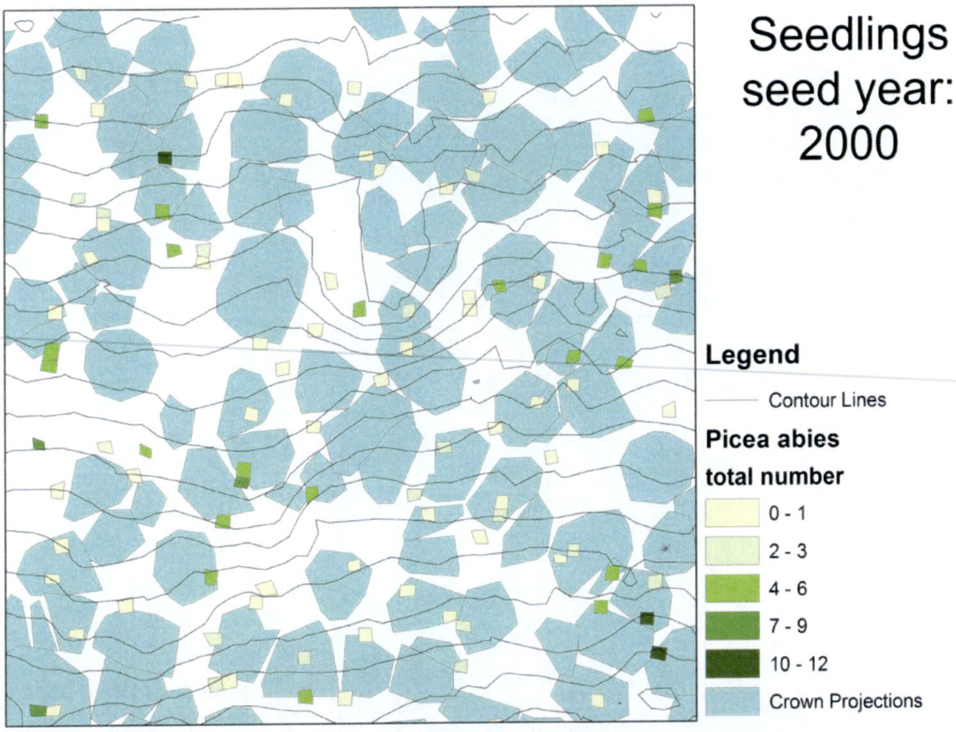

Figure 8. The map layer with tree crown projections, and the map layer with selected local vegetation sites (1x1 meter) that symbolize a total number of seedlings (*Picea abies*).

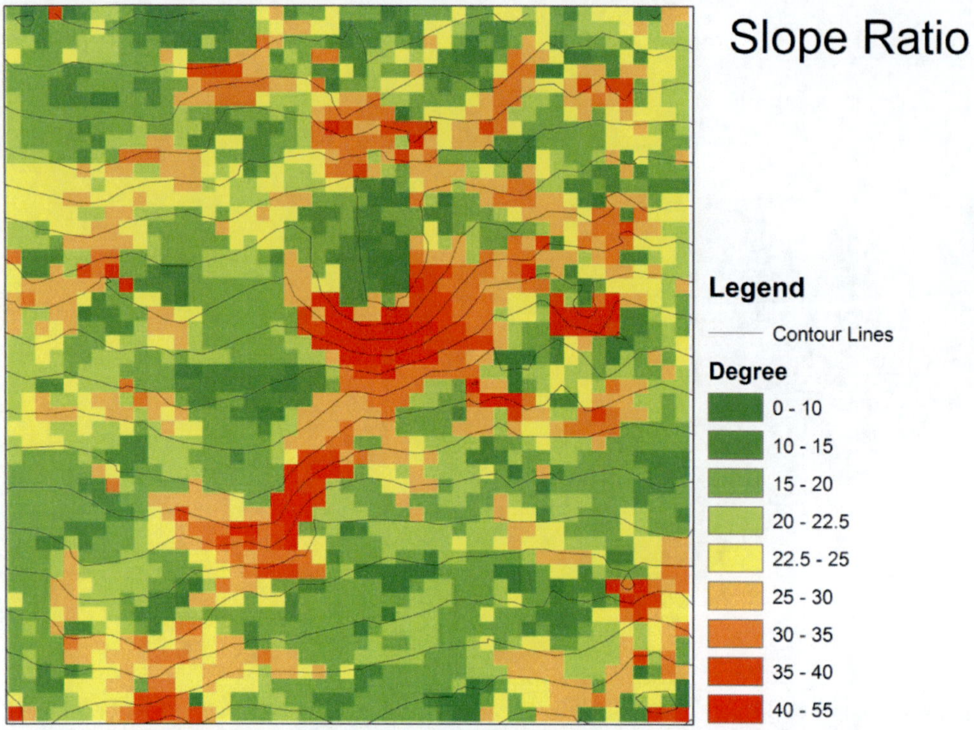

Figure 9. An example, a GIS output based on the slope ratio for setting of model parameters.

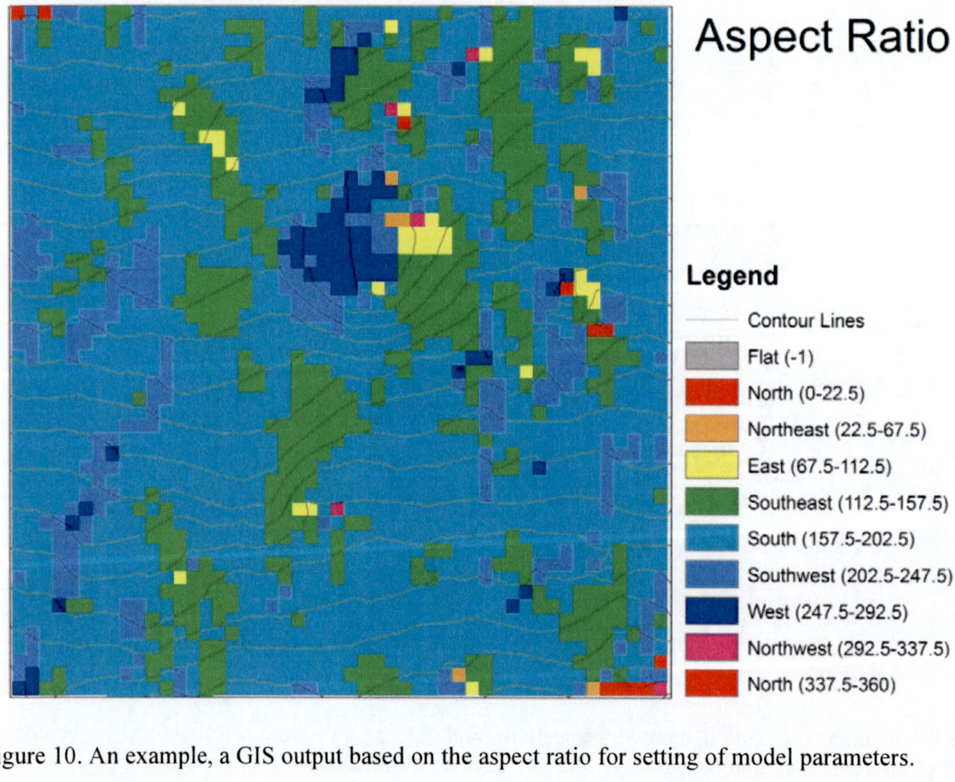

Figure 10. An example, a GIS output based on the aspect ratio for setting of model parameters.

5.3 Conclusion

Spatio-temporal modeling of environmental stress impacts on vegetation changes requires more advanced simulation tools. GISs offer a wide range of spatial methods that can be used for spatial data management, spatio-temporal analysis and visualization. A new stage is represented by ESRI's geodatabase that includes advanced spatial data structures and modeling methods. Nowadays, the dynamic modeling can be included into the GIS in order to provide simulation and data sharing in a more efficient way.

In the framework of a presented case study, the results of spatio-temporal modeling indicate that succession dynamics of floor vegetation in the declining mountain Norway spruce forests can be successfully explored, evaluated and visualized by GISs. But, more complex data will be needed to improve exploration of the local processes, and to provide more accurate predictions.

Acknowledgment

The presented research was carried out in the framework of the GIS Laboratory research project supported by the Ministry of Education, Youth and Sports. The partial results were presented at the GIS RESEARCH UK, 13th Annual Conference. The case study was based on the EU project INTAS-01-0512.

References

[1] Johnston, C.A. (1998). *Geographic Information Systems in Ecology*. London: Blackwell.

[2] Cromley, E.K. and McLafferty S.L. (2002). *GIS and Public Health*. New York: The Guilford Press.

[3] Zeiler, M. (1999). *Modeling Our World*. Redlands, California: ESRI Press.

[4] Maguire, D.J., Batty, M. and Goodchild, M.F. (2005). *GIS, Spatial Analysis and Modeling*. Redlands, California: ESRI Press.

[5] Mitchell, A. (2005). *The ESRI Guide to GIS Analysis. Volume 2: Spatial Measurements & Statistics*. Redlands, California: ESRI Press.

[6] Botkin, D.B. (1993). *Forest Dynamics*. New York: Oxford University Press.

[7] Smith, J.M. (1972). *Models in Ecology*. London: Cambridge University Press.

[8] Pielou, E.C. (1977). *Mathematical Ecology*. New York: John Wiley & Sons.

[9] Okubo, A. (1980). *Diffusion and Ecological Problems*. New York: Springer-Verlag.

[10] Murray, J.D. (1989). Mathematical biology. New York: Springer-Verlag.

[11] Odum, H.T. (1972). An energy circuit language for ecological and social systems: Its physical basis. In B.C. Patten (Ed.), *System Analysis and Simulation in ecology.* (pp. 139-211). Volume II, New York: Academic Press.

[12] Jørgensen, S.E., Patten, B.C., Straskraba, M. (1992). Ecosystems emerging: toward an ecology of complex systems in a complex future. *Ecological Modelling 62*, 1-27.

[13] Patten, B.C. (1992). Energy, emergy and environs. *Ecological Modelling 62*, 29-69.

[14] Jørgensen, S.E. (1992). Parameters, ecological constraints and exergy. *Ecological Modelling 62*, 163-170.
[15] Arctur, D. and Zeiler, M. (2004). *Designing Geodatabases: Case Studies in GIS Data Modeling*. Redlands, California: ESRI Press.
[16] MacDonald, A. (2001). *Building a Geodatabase*. Redlands, California: ESRI Press.
[17] Campbell, J.B. (2002). *Introduction to Remote Sensing*. 3rd edition. New York: Taylor & Francis.
[18] Landgrebe, D.A. (2003). *Signal Theory Methods in Multispectral Remote Sensing*. New Jersey: John Wiley and Sons.
[19] Matejicek, L. (2003). Development of Software Tools for Ecological Field Studies Using ArcPad. http://gis.esri.com/library/userconf/proc03/p0333.pdf
[20] Krebs, F. and Bossel, H. (1997). Emergent value orientation in self-organization of an animat. *Ecological Modelling 96*, 143-164.
[21] Pennanen, J., Greene, D.F., Fortin, M.J. and Messier C. (2004). Spatially explicit simulation of long-term boreal forest landscape dynamics: incorporation quantitative stand attributes. *Ecological Modelling 180*, 195-209.
[22] Bennie, J., Huntley, B., Wiltshire, A., Hill, M.O. and Baxter, R. (2008). Slope, aspect and climate: Spatially explicit and implicit models of topographic microclimate in chalk grassland. *Ecological Modelling 216*, 47-59.
[23] Skjøth, C.A., Geels, C., Hvidberg, M., Hertel, O., Brandt, J., Frohn, L.M., Hansen, K.M., Hedegaard, G.B., Christensen, J.H. and Moseholm, L. (2008). An inventory of tree species in Europe—An essential data input for air pollution modelling. *Ecological Modelling 217*, 292-304.
[24] Gudrun, W., Tappeiner, U., Strobl, J. and Tasser, E. (2008). Understanding alpine tree line dynamics: An individual-based model. *Ecological Modelling 218*, 235-246.
[25] Sohl, T. and Sayler, K. (2008). Using the FORE-SCE model to project land-cover change in the southeastern United States. *Ecological Modelling 219*, 49-65.
[26] Larocque, G.R., Archambault, L., and Delisle, C. (2006). Modelling forest succession in two southeastern Canadian mixedwood ecosystem types using the ZELIG model. *Ecological Modelling 199*, 350-362.
[27] Wernsdörfer, H., Rossi, V., Cornu, G., Oddou-Muratorio, S. and Gourlet-Fleury, S. (2008). Impact of uncertainty in tree mortality on the predictions of a tropical forest dynamics model. *Ecological Modelling 218*, 290-306.
[28] Bossel, H. (1996). TREEDYN3 forest simulation model. *Ecological Modelling 90*, 187-227.
[29] Qiu, F., Li, B., Chastain, B. and Alfarhan, M. (2008). A GIS based spatially explicit model of dispersal agent behavior. *Ecological Modelling 254*, 524-537.
[30] Rahman, M.R., Shi, Z.H. and Chongfa, C. (2009). Soil erosion hazard evaluation—An integrated use of remote sensing, GIS and statistical approaches with biophysical parameters towards management strategies. *Ecological Modelling 220*, 1724-1734.
[31] Fraser, R.H., Olthof, I. and Pouliot, D. (2009). Monitoring land cover change and ecological integrity in Canada's national parks. *Remote Sensing of Environment 113*, 1397-1409.
[32] Wulder, M.A., White, J.C., Alvarez, F., Han, T., Rogan, J. and Hawkes, B. (2009). Characterizing boreal forest wildfire with multi-temporal Landsat and LIDAR data. *Remote Sensing of Environment 113*, 1540-1555.

[33] Vacek, S., Bastl, M., Leps, J. (1999 Vegetation changes in forest of the Krkonose Mts. over a period of air pollution stress (1980-1995). *Plant Ecology 143*, 1-11.

[34] Goodchild, M.F., Steyaert, L.T. and Parks, B.O. (1996). *GIS and Environmental Modeling: Progress and Research Issues*. Fort Collins: GIS World, Inc.

[35] Burrough, P.A. (1998). *Principles of Geographical Information Systems*. Oxford: Oxford University Press.

[36] Matejicek, L., Vavrova, E., Konradova, J., Görner, T. (2006). Applications of spatial modelling in environmental studies. *Environmentalica 16*, 73-99.

In: Environmental Modeling with GIS
Editor: Lubos Matejicek, pp. 161-178
ISBN: 978-1-60876-363-4

Chapter 11

SPATIO-TEMPORAL MODELING BY FIELD COMPUTERS WITH GPS

Lubos Matejicek
Institute for Environmental Studies, Charles University in Prague, Faculty of Natural Science, Benatska 2, 128 01, Prague, Czech Republic
Jana Konradova
Institute for Environmental Studies, Charles University in Prague, Faculty of Natural Science, Benatska 2, 128 01, Prague, Czech Republic

ABSTRACT

Using field computers with GPS advanced tools can benefit from availability of a computing system, GPS data collection, and sharing data with other external sensors. It offers a complete range of properties that provide flexibility, and ease of use required by environmental mapping professionals and modelers. Using GPS and external sensors for GIS data collection and data maintenance is essential for timely decision-making and the wise use of many resources. Thus, all over the world, scientific organizations, environmental agencies, and utility companies can use GIS data collection and predictions based on model simulations to stay up-to-date and competitive. In order to demonstrate environmental mapping and spatio-temporal modeling by the field computing tools, two attached case studies have been developed to show monitoring and modeling of water quality in the small basins, and estimates in ecological terrain research. The first study focused on water quality assessment is based on the GPS high-accuracy data collection for development of the local digital terrain model, the location of hydrometric profiles, and the measurement of selected water quality indicators. The indicators are selected from the sets of general physical and chemical indicators, specific organic substances, metals and metalloids, microbiological and biological indicators, and radiological indicators. The built-in compartment models deal with basin characteristics extended by land cover attributes. Some model parameters are also estimated by experimental data and land cover types that serve as detention media or transformers. Land cover classification is provided by terrain observations, and by the spatial data derived from satellite images, and aerial photographs. Model simulations are solved by programming modules implemented in the GIS mobile software tools. Spatial analyses are based on case oriented tasks that expand the benefits of a geographic information system to field mapping efforts. All the data can be shared with the datasets managed by GIS or other information systems. The results show that the spatio-temporal modeling approach extended by the GIS and remote sensing data can support decision-making processes for better management practices in a more efficient way. The second study focused on terrain ecological exploratory analysis deals with quantitative estimates of ecological systems, such as abundance, exponential and logistic growth, migration, diversity, and similarity. Dynamic modeling formed by built-in standard ecological models is extended by spatial interactions. In addition to dynamics and spatial interactions of each population, the interaction on the level of communities or ecosystems can be subjoined. The implemented dynamic models are mostly based on the models for competition and predation. In case of spatio-temporal modeling, the diffusion and advection phenomena are implemented to simulate the migration of individuals, biomass or energy. As an example, mapping of the trees attacked by the insect species is carried out to demonstrate the mobile computing abilities. Considering to the climate conditions, the vegetation and the forest land-cover, the spreading of insect species causes large damage in the selected sites. Thus, if the attacked trees are regularly observed, the migration of the insect species can be predicted by spatial dynamic models, and consequently, compared together with the observations for the model calibration. The progress in GISs, mobile computing systems, satellite images, aerial photographs, and network interconnections represents just a part of a wide range of new approaches and technology. The presented mobile computing tools try to implement all the new phenomena in the framework of the standard surface water quality assessment, and the ecological modeling methodology.

Keywords: Spatio-temporal modeling; GPS; GIS.

1. INTRODUCTION

Advanced research progress in computing systems, satellite technology, and software development tools in a last decade have been promoted a wide range of applied scientific projects in the area of environmental science. Especially, recent mobile computing systems open new ways in data processing and measurements. There are many needs to deliver computing capabilities and data from centralized servers to a range of mobile devices in the environmental field studies. In order to manage spatial interactions together with temporal phenomena, mobile geographic information systems (GISs) are widely used for viewing, collecting, searching and synchronizing data. GISs also provide interconnections with centralized servers that can be dedicated to the central management of mobile data, digital map layers and environmental projects. Nowadays, the mobile GIS software tools for mapping offer professional GIS capabilities in the field that include field-based applications for capturing, editing, analyzing and displaying of geographic information. In addition to these capabilities, many other sensors can be linked to mobile systems, such as global positioning systems (GPS), rangefinders, digital cameras, tools for noise assessment, systems for surface water monitoring, systems for air pollution monitoring, etc. Thus, considering to a wide range of obtained data, software GIS applications are needed for mobile data processing and spatio-temporal modeling.

2. Spatio-Temporal Modeling

Environmental models are used in a vast array of GIS applications, from simple estimates to the prediction of future landscapes. In the past, the spatial software designed for high performance in dynamic modeling has been often necessary to couple with GIS. Nowadays, the increasing power of the mobile GIS hardware and software enable to solve more complex modeling tasks that cover spatial interaction and temporal phenomena. Thus, mobile applications can include simulation and validation tools. The roles of scale and accuracy, and the design of infrastructure for sharing of spatio-temporal models can be provided likewise in centralized computing tasks.

In case of GIS, the term modeling is used in several different contexts [1]. Primarily, data models represent sets of expectations about data into which the data needed for particular applications are fitted. For example, the recent geodatabase models brings physical data into objects that are mostly the same like in logical data structures, such as roads, rivers, buildings, and parcels. Further, it enables to implement the majority of custom behaviors by domains, validation rules, and other functions without writing any code [2]. Writing the software code can be only necessary for the more specialized behaviors, such as numerical simulation of temporal phenomena or case oriented spatial analysis [3]. Secondary, a model can represent processes that are believed to occur in the real world, such as mass or energy transport [4]. These models are static, if the input and the output both correspond to the same point in time, or dynamic, if the output is state dependent in addition to the input.

The power of spatio-temporal modeling sums up abilities of both the model contexts. The environmental models bring tools for description of dynamic phenomena. The data models in the framework of the GIS manage all the spatio-temporal data, which can also include pre-processing and post-processing of dynamic model inputs/outputs. Many strategies have been introduced in the books focused on GIS and environmental modeling [5].

3. Mobile Computing

Mobile GIS software tools for field mapping and data analysis have been used for a few last years. It provides field-based personnel with the ability to capture and analyze geographic information. Many professions can use mobile computers to complete a number of tasks, such as field mapping, asset inventories and maintenance, inspections and incident reporting. A key feature of GIS mobile software tools is the ability to manage a variety of geographic and attribute information, mostly, without the need to change original formats. Thus, vector and raster data can be combined in the framework of the GIS projects with the only limitations being speed and memory capacity of the hardware in use. Data models used in the centralized GIS can share data managed by mobile systems, which enables on-line checking and updating. The attributes of geographical data give an opportunity to manage descriptive information about objects, and to provide numerical calculations that include terrain estimates and dynamic modeling.

In order to capture terrain data for data models and dynamic modeling, the mobile GISs are used to be extended by GPSs, rangefinders, digital cameras and data from other sensors. For high-accuracy surface locations, differential global positioning systems (DGPS) are used together with advanced GPS technology (Trimble H-Star™ technology, post-processing based on regional reference stations). Thus, some GPS subsystems can provide real-time subfoot (<30 cm) accuracy with the internal antenna, and decimeter (10 cm) accuracy with the optional external antenna. Then, the GPS data recorded as points, polylines or polygons and their attributes are included into data models for environmental modeling. Hard-to-reach or unsafe locations can be located by rangefinders that enable measurements, such as distance, bearing or inclination. In many cases, the mobile GISs are able to manage photos from the connected camera by linking them to the actual location, and by associating them with descriptive attribute information. The basic schema of data structures and software tools for spatio-temporal modeling is in Figure 1. Spatio-temporal data in the framework of the data model are divided into four groups. The main group is formed by simulation data that include map layers with spatial objects represented by points, polylines and polygons. Each object contains attributes that can store data outputs from numerical simulation. Other groups formed by data originated from GPS, remote sensing and standard GIS layers support spatial analysis and visualization in the mobile GIS environment. The key part is represented by compartment models that contain solutions of mathematical models. In case of dynamic models, a few numerical algorithms based on Runge-Kutta methods have been implemented in the mobile GIS. All the numerical methods and other user-oriented software tools are grouped into the user programming library. Data sharing and exchange are managed by input/output scripts that can produce reports or import external data into the map layer attributes. Spatio-temporal data in the mobile GIS can be also shared by other information systems, such as GIS servers, desktop GISs or other mobile GISs.

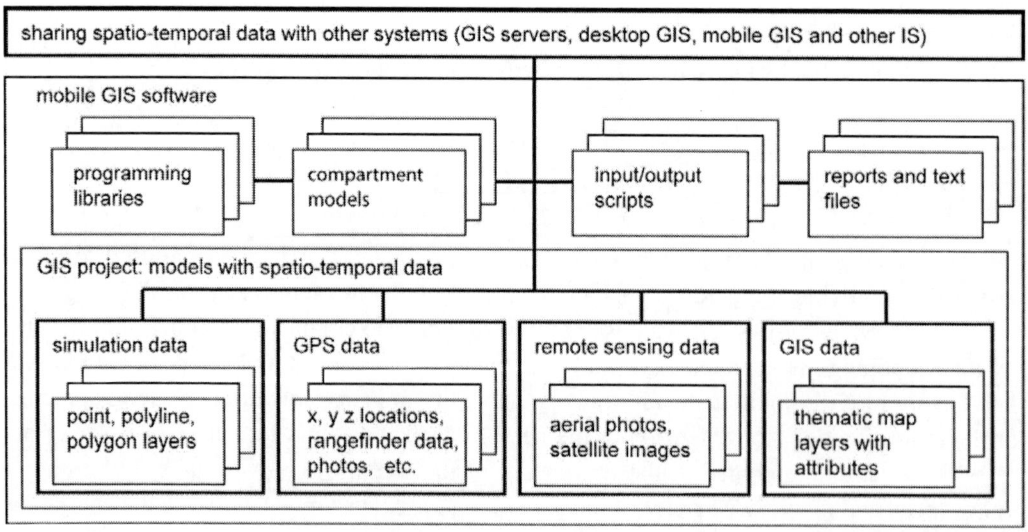

Figure 1. Structure of the mobile GIS: data models and software tools for management of spatio-temporal data based on existing data and new terrain measurements.

4. A Case Study: Monitoring and Spatio-Temporal Modeling of Water Quality in the Small Basins

As an example, a small catchment of the river Litavka is illustrated in Figure 2. It covers an area 16 km^2 of which majority is covered by forest and minority is represented by arable land and residential sites. The arable land is mostly considered as the main non-point source of surface water pollution whereas the residential sites are indicated as the point sources. The potential sources of pollution for environmental studies are formed by atmospheric deposition, and by local agricultural activities.

Spatial data accessible for field computing include thematic map layers of the catchment in a local spatial scale, aerial photos, and satellite images, optionally. The temporal data are based on a few basic surface water indicators, such as temperature, conductivity, pH and dissolved oxygen. Other indicators (nitrates, phosphates, biological oxygen demand and chemical oxygen demand) are regularly sampled and compared together with data from a whole basin. Local meteorological conditions are continuously monitored by meteorological stations with rain collectors and other sensors. All these data can be shared by field computers in order to provide complex environmental mapping and dynamic modeling. The implemented dynamic models can be used to predict surface water mass transport, such as for example, surface water pollution. In order to accelerate the numerical solution, the compartment model in Figure 3 represents the simplified modeling of the mass transport in a selected small catchment.

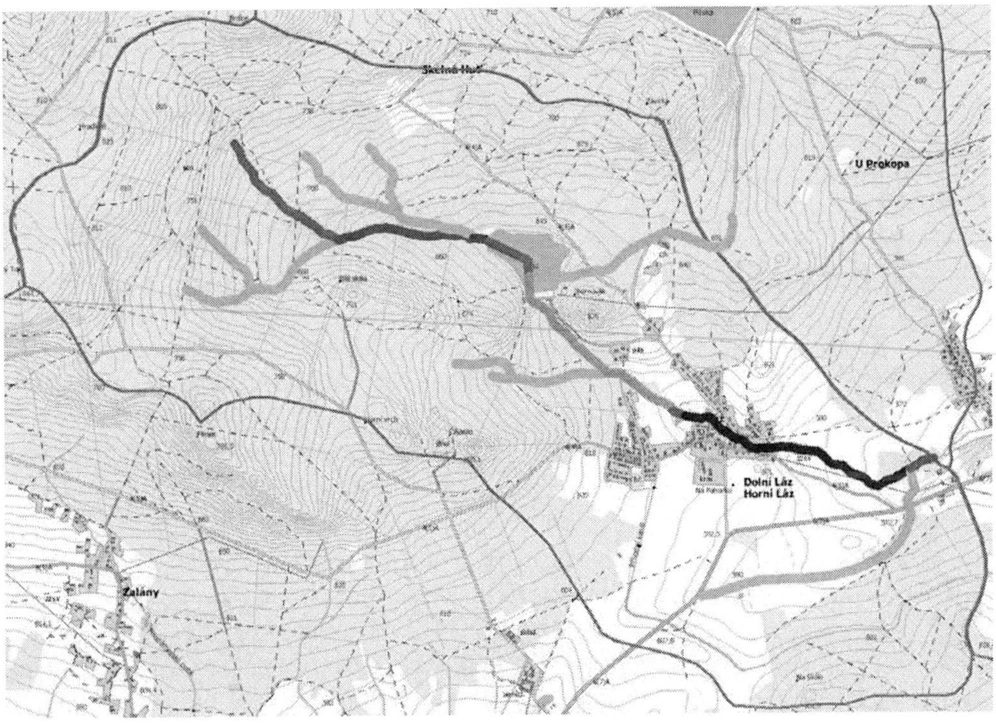

Figure 2. A small catchment of the river Litavka, an area 16 km^2 of which majority is covered by forest and minority is represented by arable land and residential sites. The stream symbols can indicate various levels of surface water pollution.

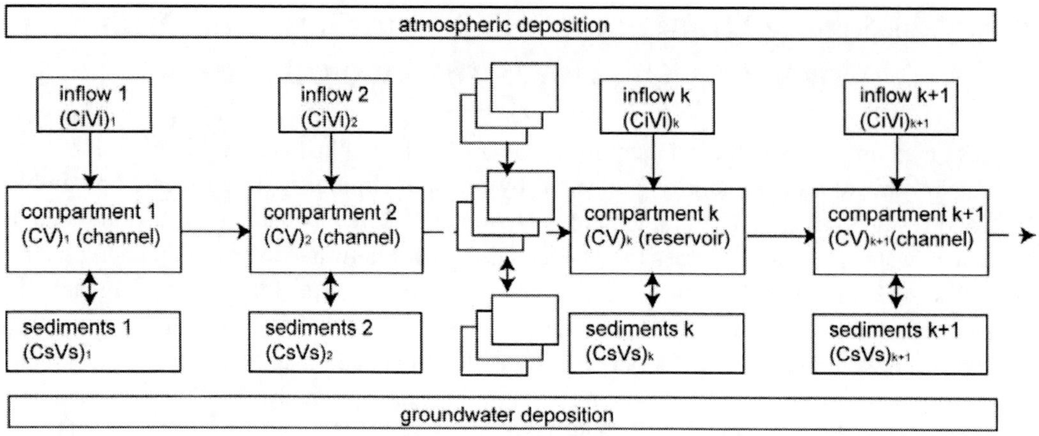

Figure 3. A basic schema of the mass transport for dynamic modeling of the surface water pollution.

4.1 Implementation of Spation-Temporal Modeling in the GIS Environment for Field Computers

The implemented compartment model in the mobile computer environment includes mass transport and accumulation in individual parts of the catchment. It can be described by an ordinary differential equation:

$$\frac{d(cv)_k}{dt} = f_{k-1} + \sum_{i=1}^{m} (f_{ki}) + f_{katm} - f_{ks} - f_{k+1}$$

where the variable $(cv)_k$ represents the pollutant amount in the kth part of the stream. The variables f_{k-1} is the pollutant inflow from the upper stream part and the variable f_{k+1} is the pollutant outflow to the downstream part. It can be replaced by the term $(cv)_k\, q_k/v_k$, where the variable q_k is the water flow and the v_k is the water volume for the kth part of the stream. The f_{ki} are the inflows and the f_{atmk} is the atmospheric deposition into the kth part of the stream. The flows between the water body and sediments are represented by f_{ks} in each part of the stream. The pollutant concentration c_k can be derived from the calculated pollutant amount $(cv)_k$ and the estimated volume v_k.

In order to provide simulation, the mathematical compartment model is transferred to the numerical model that is solved in the environment of ESRI's ArcPad. The projects created or shared by ArcPad can be customized, which requires little or no programming. For the numerical model, ArcPad exposes an extensive object model that is accessed by writing VBScripts or JScripts. This environment represents one of the mobile GISs that can assist for building of the custom applications, and for their deployment on numerous mobile devices. Numerical models based on scripts need efficient and non-time consuming calculations. Thus, the step-fixed numerical algorithms based on Runge-Kutta methods are used for solution of the ordinary differential equations by VBScripts. The mobile GISs with ArcPad also offer customized solutions that enable gathering information directly in the field, which increases the efficiency and accuracy of collecting and using information. It can include thematic maps, aerial photos and satellite images, whereas other layers support collection and management of data from terrain measurements and model simulations. In case of mass transport, such as the surface water pollutant transport, the layer attributes are used to store pollutant amounts and concentrations, water volumes, pollutant incomes from atmosphere and sediments, and water flows. Layer attributes are shared by numerical models for case oriented simulations. The simulation results are backward stored in layer attributes for display during the terrain mapping. In case of ArcPad, the spatial data stored in the map layers and located by GPS can be used directly for more complex processing in the framework of ESRI's ArcGIS projects. As an example, an implementation of the described numerical models based on ArcPad's scripts in the GeoExplorer 2008 environment is illustrated in Figure 4 that shows terrain research in a small catchment of the river Litavka.

Figure 4. Terrain research in a small catchment of the river Litavka with the field computer GeoExplorer 2008 that is used for spatial data collection by the internal GPS, storage of measurements, basic estimates, and model simulations of mass transport in the surface water network.

5. A Case Study: Development of Software Tools for Ecological Field Studies [6]

Integration of data collection, statistical analysis and dynamic modeling in ecology requires new hardware and software tools that offer mobile data management together with implementation of advanced calculation methods. Recent advances in mobile computing with ESRI's ArcPad make possible to automate a number of procedures. In addition to the mapping of spatial and temporal attributes of ecological systems, the statistical methods and dynamic models can be used to estimate new features and impacts of man-made activities. The presented software tools include parameter estimates routinely used in population ecology that are extended by spatio-temporal numerical models based on population dynamics and energy transport.

Ecology by itself is empirical science that requires data from the real world. But data are not all there is to ecology. Ecological hypotheses or ideas are the other part. Ecological field measurements depend on good sampling devices, which vary greatly for different groups of animals and plants specific to a given field of study, and appropriate statistical and modeling design. In spite of that statistical methods and modeling tools are widely used in other scientific disciplines, now they really get going in the ecological practice. It is caused by the great progress in designing of sampling devices in nearly all areas of ecology and environmental science.

In this case study, the methods for parameter estimates are divided into a few levels that are formed by ecology of the individual, population, community, and ecosystem. Data about individuals (weight or size, length or height, location, habitat and other attributes) can be collected by observers and consequently recorded directly into the digital media. Population parameters for estimates and modeling are represented mainly by the abundance or density, growth ratio, capacity and migration. Mostly, the whole population cannot be observed by itself. Thus, the various sampling techniques and statistical methods are needed, such as mark-recapture or removal methods, line intercept techniques, distance or transect methods. The dynamic modeling tools include the models for the exponential and logistic growth extended by other optional factors, such as the migration and the spatial variability. In ecology, the next hierarchical level is represented by the community that is linked with other parameters, such as similarity, diversity, interactions and niche. Again, probability and statistical techniques are used to estimates parameters of communities. In this case, dynamic models can simulate the spatial and temporal interaction. More complex ecological units are represented by ecosystems, which are mostly studied from a specific point of view, such as matter or energy flows, food chains, interaction between biotic and abiotic parts. Much of the discussion about ecological methodology is described in the book written, for example, by Krebs [7]. Statistical theory and ecological applications implemented in the software tools come out from the books written by Legendre [8], Sokal [9], and others [10, 11]. Examples of ecological dynamic models are derived from Pielou [12] and Krebs [13]. Some dynamic models are extended by spatial aspects that are discussed, for example, by Okubo [14] and Shigesada [15].

5.1 Implementation of Ecological Methods in the GIS Environment for Field Computers

The software tools for ecological analysis are developed with the ESRI's ArcPad Studio, which assists in the design of user data entry forms and scripts. All the applications run in the ESRI's ArcPad that can be also installed in the mobile computing environment. The ArcPad is a GIS's product for mobile computing that supports a multi-layer environment with the industry-standard vector map and the raster image layers. Its ability to work directly in the shape formats with symbology exported from ESRI's ArcView or ArcGIS, and the interface with the GPS makes it useful in a number of field ecological data collections. The software tools are developed with VBScript language. The ArcPad user interface is customized by files based on the ArcPad's XML format. All these tools are sufficient for calculation of estimated ecological parameters, and simulation of smaller spatial dynamic models. The spatial data exchange in the framework of the GIS's data formats enable to move some larger time consuming calculations to the ESRI's ArcGIS. Consequently, the computations can be carried out by extensions developed with the use of ESRI's ArcObjects. The numeric algorithms included in the multidimensional data analysis and spatio-temporal modeling are discussed in the book written, for example, by Flowers [16].

The software tools include a number of methods focused on spatial data management, multidimensional data analysis and dynamic modeling. The basic schema of the mobile system is in Figure 5, which illustrates its basic interoperability.

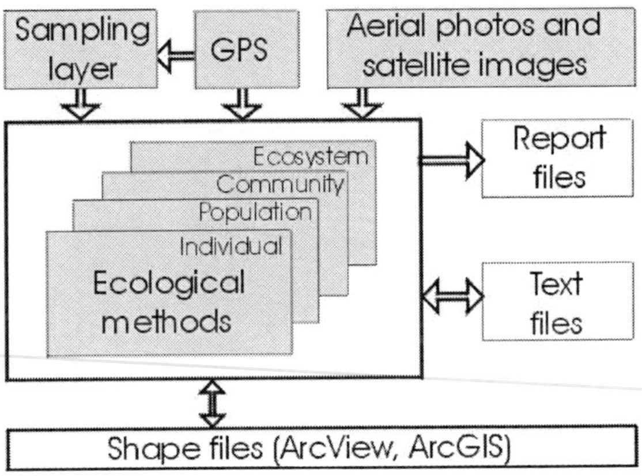

Figure 5. The general schema for implementation of the described calculation tools for mobile computing systems.

The sampling layer is used for collection of the individual samples. Data are collected into the three forms, Figure 6. The first form contains identification codes, date and time of an observation or capture, a name of the specie and its habitat, a name of the observer and the location based on coordinates (X, Y, Z). The second form is used for input values that can represent variables in the nominal, ordinal, interval or ratio statistical scales. Their meaning depends on each practical case. The third form is completed by other information, such as comments and images. The coordinates can be captured via the GPS connected to the mobile computer, by placing the pen on the map layer or by typing appropriate coordinates into the form. All the data are stored as attributes of the sampling layer.

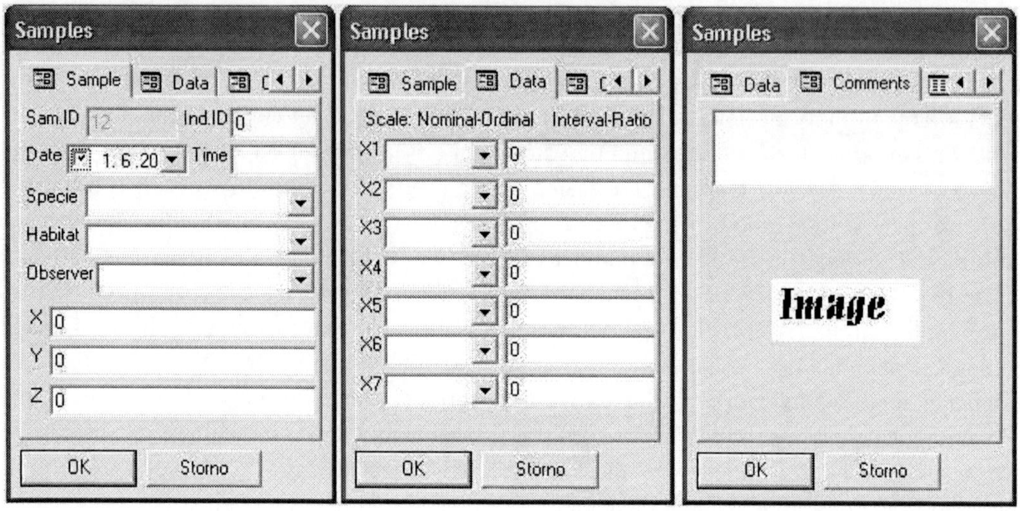

Figure 6. User forms of the sampling layer that includes three parts: inputs for basic information, advanced inputs for multidimensional data analysis and dynamic modeling, and other information (comments or images).

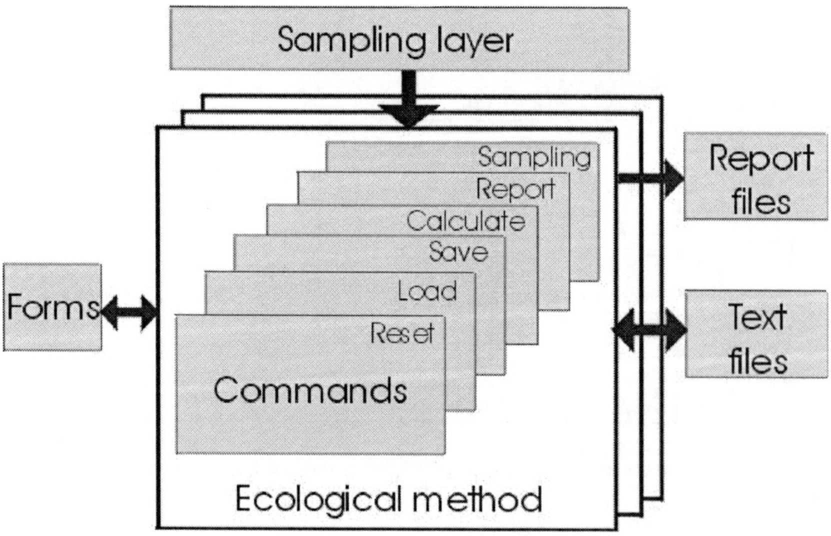

Figure 7. Basic schema of the data management for each ecological method that is implemented in the mobile computing environment.

In addition to the sampling layer, other vector and raster layers (thematic maps, aerial and satellite images) are usually used to complete the project, and to help the observer to identify the position and the right place for the observation. Consequently, data from the sampling layer are transferred into the modules that provide calculations and estimates based on ecological methods. Each method has its own form and the data format for saving of the input and output data. The input data can be also placed directly into the form and stored in the file without the transfer from the sampling layer. Even though the data are saved in the text format, the report files can be built to show the input data and the results in a more user friendly way. The basic data management for each implemented ecological method is illustrated in Figure 7.

Generally, the *Reset* command sets the initial values of the inputs. The *Load* and *Save* commands are used for the data store. The *Calculation* command starts the computing. The input and output data can be stored by the *Report* command into the text file. The *Sampling* command transfers the data from the sampling layer into the form of a selected ecological method. As an example, the forms for three ecological methods (estimates of the population abundance with the mark-recapture techniques) are presented in Figure 8. The button *SAMPLE* is used to transfer data from the selected layer, such as sampling layer or another layer. The next part of the form contains the input menu. The bottom part of the form is reserved for the data outputs that are set after the calculation. The buttons of individual commands are placed in the middle part of the forms.

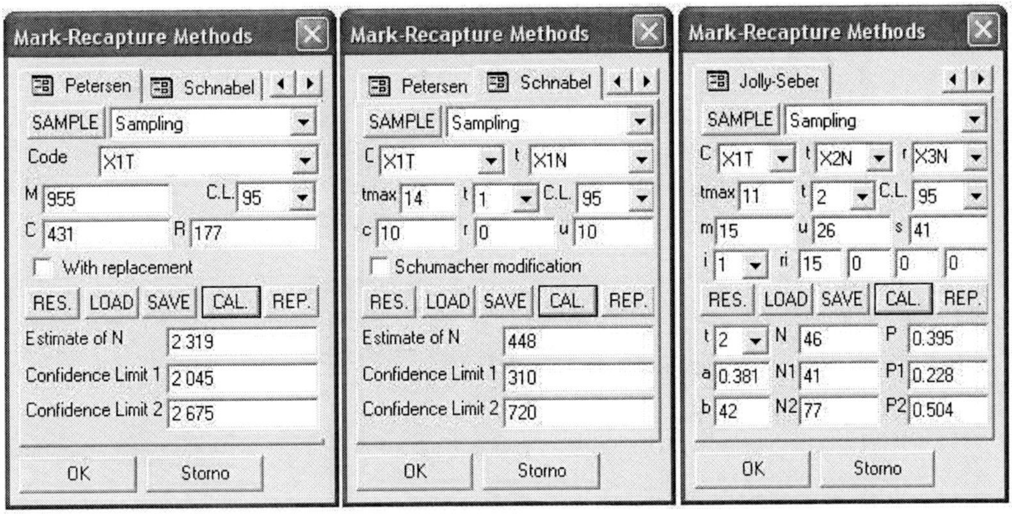

Figure 8. Examples of the forms for the mark-recapture techniques: estimates described by Petersen, Schnabel-Schumacher and Jolly-Seber (tested and processed with the data from Krebs [13]).

Each ecological method is grouped into the appropriate toolbar. In addition to the standard toolbars, the toolbars for data capture from the sampling layer, from the GPS and via the network from the GIS servers are added to make methods accessible for the user. All the ecological methods are placed into two bottom toolbar groups that are mostly concerned on methods dedicated for exploration of the population level and the community level, Figure 9. In reality, each button by itself represents a class of ecological methods that is focused on the specific ecological estimates or the selected model solutions.

Figure 9. User toolbars in the ArcPad environment.

5.2 Implemented Ecological Methods for Analysis and Modeling of Population

There is a wide range of methods in ecology. They are used for estimates of ecological parameters, calculations of empirical rules, and simulations of ecological models. In this part, the main attention is focused on description of the methods for population ecology. Due to the speed and the capacity of the present mobile computers, the lager dynamic models tend to be simulated in the standard computing environment. The selected methods include mark-recapture techniques, removal methods, quadrat counts, line transects and distance methods that are extended by estimates of the mortality, migration and growth ratio.

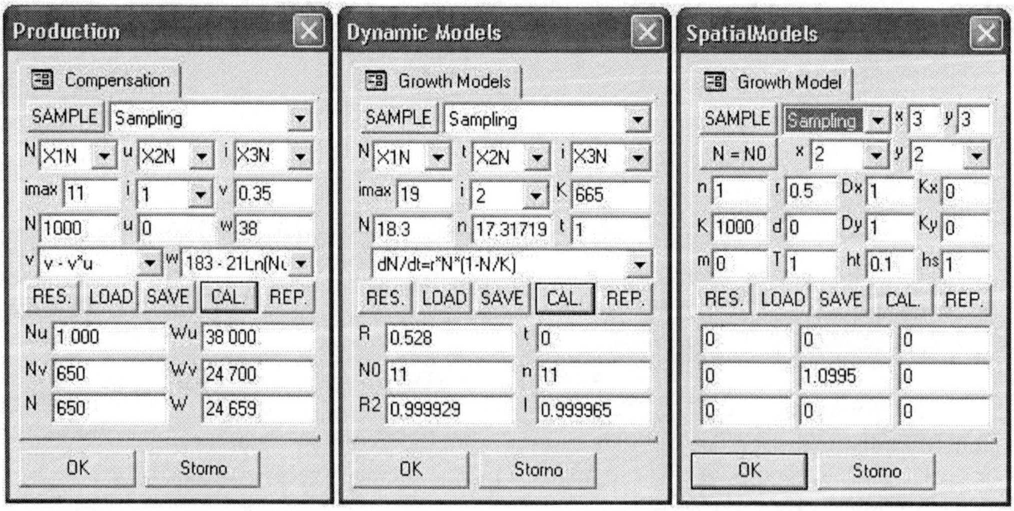

Figure 10. Examples dedicated to the biomass compensation, the growth models, and the spatio-temporal growth models.

The mark-recapture techniques include the methods described by Petersen, Schnabel and Schumacher for the closed population, and the method made out by Jolly and Seber for the opened population. In addition to the estimates of the abundance and related parameters, the confidence limits enable to assess the variability of the results. Other alternatives to the mark-recapture techniques represent removal and resight methods, quadrat counts, line transect and distance methods. The methods implemented in the ArcPad include the catch-effort method (Leslie-Ricker), the line intercept technique and the Hayne estimator. The distance methods are represented by the Byth and Ripley procedure, the T-square sampling procedure and the ordered distance method. Mostly, the variability of the estimates is carried out by confidence limits on a few levels of critical values [7]. Other class of methods is concerned on spatio-temporal modeling and optimization. Among a wide range of various methods and techniques, the growth models, such as the exponential and the logistic growth, together with estimates of their parameters, are implemented in the ArcPad, Figure 10. The dynamic models formed by the logistic growth can be extended by spatial interaction. Their spatial dataset is represented by the regular grid of cells. Each cell contains as its property the abundance and other attributes, such as the initial value, growth ratio, capacity, and migration ratio. All the data can be directly shared and visualized in the map layers in the ArcPad or transferred into to another GIS, such as ESRI's ArcGIS.

5.3 Implemented Ecological Methods for Analysis and Modeling of Comunities

Attempting to characterize and measure the properties of ecological communities faces a special set of statistical and modeling problems. Some of these methods and techniques are also implemented in the ArcPad. They are mostly focused on the evaluation of similarity parameters, cluster analyses and multivariate statistical techniques, diversity measures, niche characteristics, and dynamic models extended by spatial interaction. The classification of communities or their samples needs measures of similarity. The various evaluation methods of similarity have been carried out to compare communities in the best way. Also, much confusion exists about which measure to use. There are two classes of similarity measures implemented in the ArcPad. The binary measures contain the Jaccard coefficient, the simple matching coefficient, the Sorensen coefficient and the Baroni-Urbani, and Buser coefficient. The distance measures are represented by the Euclidean distance, the Bray-Curtis measure, the Canberra metric, the percentage similarity, the Morisita's index, and the Horn's index of similarity. In cases of many samples of the communities to analyze, various techniques for grouping of the similar samples are used. The implemented methods include a few clustering methods such as the single, complete and average linkage clustering. The user forms of the mentioned methods are illustrated in Figure 11.

Species diversity is evaluated by many different ways. Among the various concepts of diversity, the species richness and heterogeneity are widely used. The computing system contains calculation of the species richness that is based on the rarefaction method. Evaluation of the species heterogeneity is carried out with the Simpson's index, the Shannon-Wiener function and the Brillouin's index. Measurements of the niche parameters of a population to analyze community dynamics can be provided by a number of techniques. The implemented methods are divided into the evaluation of the niche breadth, and the niche overlap. The niche breadth, which is also called the niche width or the niche size, is calculated by the Levins's measure, the Shannon-Wiener measure and the Smith's measure. The calculation of the niche overlap is represented by the MacArthur and Levins's measure, the percentage overlap, the Morisita's measure, the simplified Morisita index, the Horn's index, and the Hurlbert's index. A group of the multidimensional data analysis contains a number of more general methods that are represented, for example, by the principal component analysis (PCA), the multiple correlation analysis, and the multiple regression analysis, Figure 12. In spite of that the methods are mostly implemented in statistical software, the ArcPad environment enables their integration in the framework of the basic field calculations, ecological estimates, and spatio-temporal modeling. Another class of the statistical methods is formed by hypothesis testing. In the ArcPad, a number of extensions can be established to extend standard ecological methods.

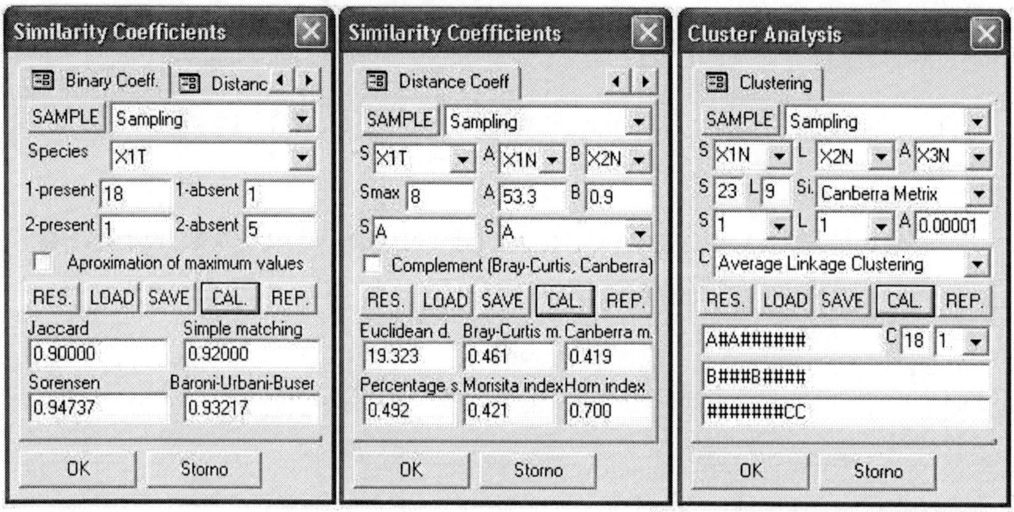

Figure 11. Similarity coefficients and clustering: the binary and distance coefficients, clustering (tested and processed by data from Krebs [13]).

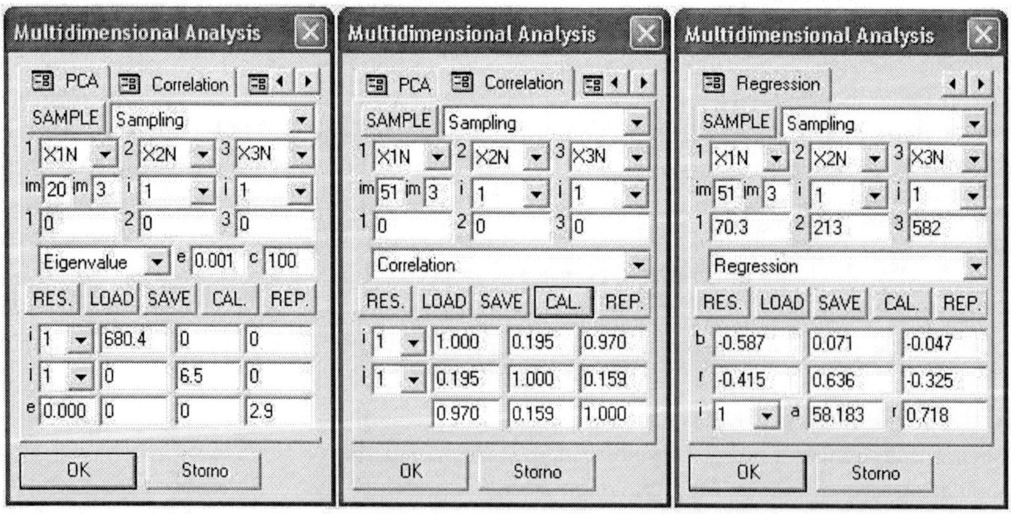

Figure 12. The multidimensional data analyses: the regression and correlation analysis, the ordination techniques (tested and processed by data reported in the framework of the research projects).

Dynamic modeling represented by the standard ecological models, which describe community dynamics and basic types of interactions, is extended by spatial interaction. In addition to the dynamics and spatial interaction of each population, the interactions on the level of communities or ecosystems can be subjoined. The implemented dynamic models are mostly based on ecological competition models. In case of spatio-temporal modeling, the diffusion and advection phenomena are implemented to simulate the migration of individuals, biomass or energy. Due to the amount of data calculated in each time step, only the last data grid is stored into the data output. Likewise in the spatial modeling of population, the larger models are planned to be computed and visualized with a more powerful computer equipment.

Figure 13. Spreading of the insect species in the forest cover: the simulation study carried out by the ArcPad built-in spatio-temporal modeling tools.

As an example, mapping of the trees attacked by the insectspecies is carried out. Considering to the climate conditions and a type of the forest cover, the spreading of the insect species causes large damage. If the attacked trees are regularly observed, the migration of the insect species can be predicted by spatio-temporal models, and consequently, compared together with the terrain observations. The input into the dynamic model is represented by the aerial image and the terrain sampling. The basic tools enable efficient calculations of smaller data grids. The outputs of the models are formed by individual simulation studies that contain time series of the insect species densities in the data grids. The results can be displayed and symbolized in the GIS environment. The attached example shows integration of the data based on spatio-temporal simulation in the mobile GIS together with the data from remote sensing, and the GPS, Figure 13.

6. Conclusion

Using next-generation rugged handheld computing devices with GPS technology is setting new mapping and GIS industry standards with real-time decimeter level positioning accuracy, sharing data with external sensors and mobile GISs that best meet specific field requirements. Demonstrated mobile GISs offer a customized solution in order to manage digital map layers, aerial photos, satellite images, data from GPS and external sensors. Its programming capabilities enable to create a number of analytical functions and modeling tools that are best fitted to the area of interest [17, 18]. Thus, many projects are going to be developed in the framework of integration environmental research with mobile GISs complemented by GPS and remote sensing.

Traditionally, the processes of field data collection and editing have been time consuming and error prone. Data collections were based on notes and sketches on forms and paper maps that were deciphered and manually entered into GIS databases. Developments in the mobile GIS have enabled the GIS databases to be taken into the field as digital maps providing field access to enterprise geographic information. It offers terrain researches to add real-time information to their database and modeling applications, speeding up analysis, display, and decision making by using actual, more accurate spatio-temporal data. Thus, the mobile GIS can support field mapping, asset inventories, inspections, incident reporting and a wide range of GIS methods recently processed in the desktop GIS environment [19, 20].

The attached case studies are used to document large capabilities of the mobile information systems, especially GISs, to process spatio-temporal data with the field computers. In spite of that the recent mobile GISs are mainly used for integration of data sources based on monitoring, the feature powerful mobile computing systems and accurate sensors will be able to provide more complex data analysis and spatio-temporal modeling tasks.

Acknowledgments

The first study was supported in part by the project of the Czech Science Foundation (GACR) 526/09/0567. Some digital thematic maps used for environmental terrain studies are administrated by CENIA in the Czech Republic (http://geoportal.cenia.cz, 2009/05/01). The compartment model was implemented in the mobile GIS (ESRI's ArcPad 8.x, evaluation mode, TRIMBLE-GeoExplorer 2008) in the framework of the GIS and environmental modeling lectures supported by the Ministry of Education Youth and Sports of the Czech

Republic in 2009. The final documents were processed by ArcGIS and ERDAS Imagine in the GIS Laboratory, Charles University in Prague.

The second study was based on the presentation of the EcoPad software at the 23rd ESRI International Conference in San Diego that was partly supported by the research project MSM113100007. The EcoPad (version 1.0) was developed within the scope of the research project MSMT97100 in the framework of the Campus Pack license from the Environmental Systems Research Institute (ESRI).

REFERENCES

[1] Goodchild, M.F. (2005). GIS, spatial analysis, and modeling overview. In: D.J. Maguire, M., Batty, M.F., Goodchild (Eds), *GIS, Spatial Analysis, and Modeling*. Redlands, California: ESRI Press.

[2] Zeiler, M. (1999). *Modeling Our World*. Redlands, California: ESRI Press.

[3] Matejicek, L. (2006). Modeling of water pollution in river basins with GIS. In: A.R. Burk (Ed). *Focus on ecology research*. New York: Nova Science Publishers.

[4] Parker, P., Letcher, R., Jakeman, A., Beck, M.B., Harris, G. (2002). Progress in integrated assessment and modelling. *Environmental Modelling & Software 17*, 209-217.

[5] Goodchild, M.F., Styaert, L.T., Parks, B.O. (1996). *GIS and Environmental Modeling: Progress and Research Issues*. Fort Collins: GIS World, Inc.

[6] Matejicek, L. (2003). *Development of software tools for ecological field studies using ArcPad*. San Diego: ESRI International Conference 2003.

[7] Krebs, C.,J. (1999). *Ecological methodology*. New York: Addison Wesley Longman.

[8] Legendre, P. and Legendre, L. (1998). *Numerical ecology*. Amsterdam: Elsevier.

[9] Sokal, R.R. and Rohlf, F.J. (1995). *Biometry: The principles and practice of statistics in biological research*. New York: W.H. Freeman.

[10] Heath, D. (1995). *An introduction to experimental design and statistics for biology*. London: UCL Press.

[11] Kanji, G.K. (1993). *100 Statistical tests*. London: SAGE.

[12] Pielou, E.C. (1977). *Mathematical ecology*. New York: Wiley.

[13] Krebs, C.,J. (1994). *Ecology: The experimental analysis of distribution and abundance*. New York: Harper Collins.

[14] Okubo, A. (1980). *Diffusion and ecological problems: Mathematical models*. Berlin: Springer.

[15] Shigesada, N. and Kawasaki, K. (1997). *Biological invasions: Theory and practice*. Oxford: Oxford University Press.

[16] Flowers, B.H. (1995). *An introduction to numerical methods in C++*. Oxford: Clarendon Press.

[17] Mitchell, A. (2005). *The ESRI Guide to GIS Analysis. Volume 1: Geographic Patterns and Relationships*. Redlands, California: ESRI Press.

[18] Mitchell, A. (2005). *The ESRI Guide to GIS Analysis. Volume 2: Spatial Measurements & Statistics*. Redlands, California: ESRI Press.

[19] Argent, R.M., Perraud, J.M., Rahman, J.M., Grayson, R.B. and Podger G.M. (2009). A new approach to water quality modelling and environmental decision support systems. *Environmental Modelling & Software 24*, 809-818.

[20] Li, H., Chen, Z., Yong, L., Kong, S.C.W. (2005). Application of integrated GPS and GIS technology for reducing construction waste and improving construction efficiency. *Automation in Construction 14*, 323-331.

INDEX

A

B

D

E

F

G

H

I

J

K

L

M

Q

R

S

T

U

V

W

X